ENERGY POLICIES OF IEA COUNTRIES

PORTUGAL 2004

CORRIGENDUM for FIGURE 25, p 88

The legend concerning Figure 25 has been inadvertently omitted.
The following provides the complete version of this figure.
We apologise for any inconvenience caused.

Figure 25

Electricity Prices for Industry in Portugal and Spain, 1 July 2003

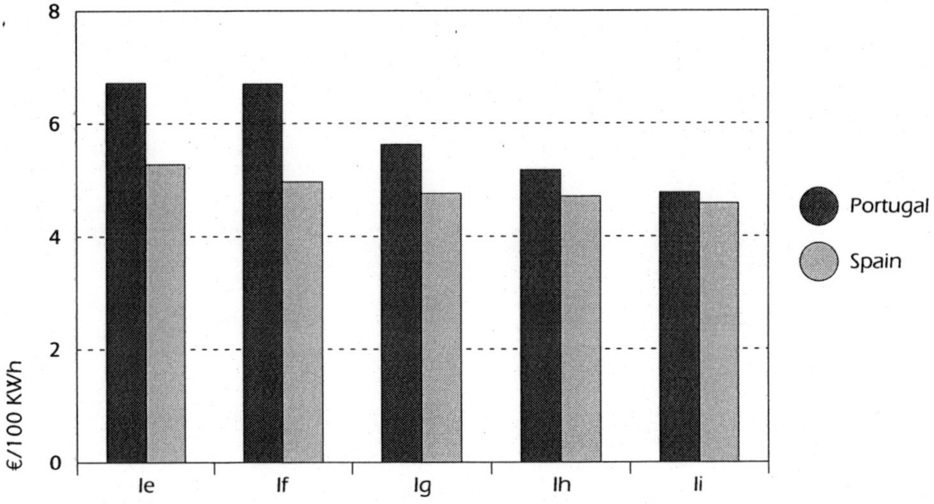

Standard consumer	*Annual consumption (kWh)*	*Maximum demand (kW)*	*Annual utilisation (in hours)*
Ie	2 000 000	500	4 000
If	10 000 000	2 500	4 000
Ig	24 000 000	4 000	6 000
Ih	50 000 000	10 000	5 000
Ii	70 000 000	10 000	7 000

Source: Eurostat.

INTERNATIONAL ENERGY AGENCY

Energy Policies of IEA Countries

PORTUGAL

2004 Review

INTERNATIONAL ENERGY AGENCY
9, rue de la Fédération,
75739 Paris Cedex 15, France

The International Energy Agency (IEA) is an autonomous body which was established in November 1974 within the framework of the Organisation for Economic Co-operation and Development (OECD) to implement an international energy programme.

It carries out a comprehensive programme of energy co-operation among twenty-six* of the OECD's thirty member countries. The basic aims of the IEA are:

- to maintain and improve systems for coping with oil supply disruptions;
- to promote rational energy policies in a global context through co-operative relations with non-member countries, industry and international organisations;
- to operate a permanent information system on the international oil market;
- to improve the world's energy supply and demand structure by developing alternative energy sources and increasing the efficiency of energy use;
- to assist in the integration of environmental and energy policies.

** IEA member countries: Australia, Austria, Belgium, Canada, the Czech Republic, Denmark, Finland, France, Germany, Greece, Hungary, Ireland, Italy, Japan, the Republic of Korea, Luxembourg, the Netherlands, New Zealand, Norway, Portugal, Spain, Sweden, Switzerland, Turkey, the United Kingdom, the United States. The European Commission also takes part in the work of the IEA.*

ORGANISATION FOR ECONOMIC CO-OPERATION AND DEVELOPMENT

Pursuant to Article 1 of the Convention signed in Paris on 14th December 1960, and which came into force on 30th September 1961, the Organisation for Economic Co-operation and Development (OECD) shall promote policies designed:

- to achieve the highest sustainable economic growth and employment and a rising standard of living in member countries, while maintaining financial stability, and thus to contribute to the development of the world economy;
- to contribute to sound economic expansion in member as well as non-member countries in the process of economic development; and
- to contribute to the expansion of world trade on a multilateral, non-discriminatory basis in accordance with international obligations.

The original member countries of the OECD are Austria, Belgium, Canada, Denmark, France, Germany, Greece, Iceland, Ireland, Italy, Luxembourg, the Netherlands, Norway, Portugal, Spain, Sweden, Switzerland, Turkey, the United Kingdom and the United States. The following countries became members subsequently through accession at the dates indicated hereafter: Japan (28th April 1964), Finland (28th January 1969), Australia (7th June 1971), New Zealand (29th May 1973), Mexico (18th May 1994), the Czech Republic (21st December 1995), Hungary (7th May 1996), Poland (22nd November 1996), the Republic of Korea (12th December 1996) and Slovakia (28th September 2000). The Commission of the European Communities takes part in the work of the OECD (Article 13 of the OECD Convention).

TABLE OF CONTENTS

Tables and Figures

TABLES

FIGURES

SUMMARY OF CONCLUSIONS AND RECOMMENDATIONS

SUMMARY

Since the last IEA in-depth review in 2000, there have been many positive developments in Portugal's energy sector. The Portuguese government decided in 2003 to restructure the energy sector. It set guidelines for energy policy associated with a demanding implementation plan. Natural gas is successfully penetrating the energy mix. Important political decisions have been taken to develop the Iberian electricity market. It has also been decided to fully liberalise the electricity market from July 2004. Oil prices have been totally liberalised. The regulator (ERSE) is functioning well with high independence and strong authority. A national action strategy to tackle climate change has been proposed and renewables are given high priority. Portugal is making efforts to tackle the many challenges in the years to come.

SECURITY OF SUPPLY

The share of oil in Portugal's TPES was 62% in 2002 and has been decreasing over the last decade, but it is clearly higher than the IEA average of 41% (2001). Consequently, oil supply is very important for Portugal's security of supply. However, Portugal has frequently been non-compliant with its IEA obligation of holding oil stocks equivalent to 90 days of net imports since late 1992. This situation has been continuously addressed by the IEA and its members, as energy security in today's energy environment depends on solidarity. Positive decisions have been taken by the government but they still need to be fully implemented.

The government of Portugal should be commended for its efforts in the introduction of natural gas into the energy mix. Natural gas will reduce Portugal's high dependence on imported oil and will continue to diversify energy supply. While Portugal had been dependent on a single supply source (Algeria), the new liquefied natural gas (LNG) terminal has significantly enhanced security of supply both by allowing diversification of supply sources and by providing enough supply capacity for several years. Also the new underground storage facility enhances security of supply.

ENERGY AND THE ENVIRONMENT

The EU Burden-Sharing Agreement under the Kyoto Protocol commits Portugal to limit its increase of greenhouse gas (GHG) emissions to 27% between 1990 and the first commitment period of 2008-2012. Since 1990 to 2001, GHG emissions grew by 36.4%, in line with GDP growth, but already 9.4 percentage points over the target. Even with the policies and measures in place, emissions are expected to stay above the final target. Therefore, meeting it will be a challenging task. Prompt introduction of new, strong policies and measures is required. Furthermore, the baseline estimates should be reviewed when necessary to reflect the rapid growth in energy demand, in particular in the transport sector.

The introduction of the Climate Change National Programme (PNAC) will be particularly important and should include effective policies and measures with a clear implementation plan, time schedules as well as adequate funds for implementation. Careful monitoring of the trends and the effectiveness of policies, followed by a prompt policy response, is important. More attention needs to be paid to the cost-effectiveness of the policies and measures. Introduction of the EU Emissions Trading Scheme (EU-ETS) will be an important device to curb GHG emissions from the large emitters. Because the reference scenario shows that domestic measures are not enough to meet the target, the use of the Kyoto flexible mechanisms will be necessary. This is a sensible policy as the use of these mechanisms is likely to reduce the cost of achieving the target.

END-USE EFFICIENCY

It is a matter of concern that since 1990, energy demand in Portugal has grown faster than GDP, thus increasing energy intensity. However, energy intensity remains below the IEA average, but the growing trend is in contrast with the trend in most other IEA member countries. While this is partly attributable to the improving living standards, there is potential in Portugal to improve energy efficiency. Better energy efficiency would also contribute positively towards energy security. Though significant new action has been taken recently to strengthen the policies, a more solid energy efficiency strategy is necessary. This should include measurable targets and thorough monitoring. The policies and measures to be included should be chosen on the basis of their cost-effectiveness.

The introduction of the EU-ETS could give a good incentive to large emitters to further improve their energy efficiency, but a new type of arrangement would be necessary for emitters not covered by the EU-ETS. Policies to de-link transport demand growth from economic growth need to be thoroughly explored and the process should be supported by effective co-ordination among different ministries.

RENEWABLES

Portugal has a demanding indicative target to increase the share of renewables in electricity generation to 39% by 2010, and specific targets for several technologies. For this purpose, the government has relied on a feed-in tariff scheme which has the same impact as subsidies. Additional efforts are needed so that the required extra capacity will be built in time. The government should also seek to improve the current system so that the final consumers could benefit from the cost reduction through gradually reducing the feed-in tariff and limiting the duration of the buy-back period while ensuring investor confidence.

GAS MARKET REFORM

The new challenge in the gas sector is market liberalisation and the announced intention to create an Iberian gas market. The incumbent supplier, GALP Energia, dominates the gas market from upstream to retail. It is commendable that the government has announced its intention to bring forward the beginning of the gas sector liberalisation from 2008 to 1 July 2004, even though the country has a derogation from the EU directive. Following this move, it is important to implement as soon as possible a regulatory framework which is clear, gives sufficient powers to the regulator and provides incentives for an efficient market. The third-party access tariff for networks needs to be set and should provide for fair, non-discriminatory and transparent access. One prerequisite for implementing these principles is effective unbundling of the supply and network operations.

ELECTRICITY MARKET REFORM

The electricity market in Portugal has developed significantly since the last in-depth review, but remains dominated by Electricidade de Portugal (EDP). However, there are several initiatives to induce electricity market competition in Portugal and at the Iberian level. Increasing physical interconnections with Spain would improve competition and security of supply. The development of the natural gas network has increased the feasibility of independent entry into power generation in Portugal. Particularly, the political initiative to create a common Iberian electricity market is an important step in creating a competitive market but progress is behind schedule. The creation of the Iberian market requires adequate legislation in both countries to make them compatible but it is yet to be detailed and published. Furthermore, close monitoring of the prices and bidding behaviour of the major players in the Iberian market, particularly in terms of wholesale prices in Portugal, will be necessary.

The need to unwind the long-term power purchase agreements (CAEs) between Rede Eléctrica Nacional (REN) and main suppliers creates an important opportunity to enhance competition in the Portuguese market. One possible way would be auctioning the contractual obligations by REN to new suppliers, which would make more electricity available to new entrants. This would have the impact of reducing EDP's market share until a better interconnection can be achieved. However, the issue of concentrated generation in the Iberian electricity market can only really be dealt with at the Iberian level.

On 1 July 2004, Portugal plans to open the electricity market to all consumers in the country, as already exists in Spain. However, it is still uncertain that meaningful competition will develop quickly. Legislation and regulations will be needed to clarify the rules for small consumers to switch suppliers. While it is proposed to eliminate the regulated tariff option for all consumers, this process should take place at the Iberian level, starting with the largest consumers and allowing for the existence of suppliers of last resort.

A policy of capacity payments, which places obligations on retailers to acquire more than sufficient capacity to supply consumers, has not been proven to be an effective or efficient policy to secure adequate supply. More efficient instruments should be considered. Enhancing the demand response of consumers to the market price of electricity would reduce the need to invest in new generating capacity while increasing the efficiency of the market and reducing price volatility.

RESEARCH AND DEVELOPMENT

Given that Portugal is facing significant energy and environment policy challenges, the government needs to explore all possible means to respond to these challenges, including formulating an effective energy R&D policy. To implement such a policy, a coherent energy R&D strategy with adequate financing, as well as better co-operation between the different ministries and the research laboratories under them, are necessary. To increase the cost-effectiveness of energy R&D and to make better use of the results, it is necessary to improve monitoring and assessment mechanisms.

RECOMMENDATIONS

The government of Portugal should:

General Energy Policy

- *Continue efforts to diversify the energy mix and closely monitor progress to improve security of supply and enhance climate change mitigation and environmental protection.*

- *Enhance co-ordination of energy policy measures between the different ministries and appropriate organisations to take better account of energy in other policies and to increase efficiency.*

- *Involve all stakeholders, including consumers, in developing energy policies and disseminate information to them.*

- *Continue the close co-operation with the government of Spain to develop consistent policies for the Iberian energy markets. This includes electricity tariff schemes, the access and availability of natural gas, the allocation of CO_2 emission permits, and the policies regarding renewables and security of supply.*

- *Scrutinise the impacts of EDP's acquisition of GALP's gas supply and retailing activities on competition in the electricity and natural gas markets.*

- *Make sure that ERSE is given the necessary means to conduct its duties, also in the gas sector.*

Energy and the Environment

- *Finalise the Climate Change National Programme (PNAC) and adopt a time schedule for the policies and measures with reasonable and achievable goals at each step. Periodically review on the basis of recent GHG projections and the progress of various policies and measures.*

- *Further examine the cost-effectiveness of climate change policies and measures.*

- *Clarify the role which Kyoto flexible mechanisms should have in meeting the emissions targets and accelerate the preparation for their utilisation.*

- *Increase the use of fiscal instruments to internalise the externalities related to climate change and air pollution.*

Energy Efficiency

- *Establish an integrated national energy efficiency strategy as soon as possible. This should incorporate, at the national and sectoral levels, targets and strong cost-effective measures, including full implementation of the measures in the EU framework.*

- *Evaluate the cost-effectiveness and achieved energy savings of the energy efficiency projects.*

- *Enhance efforts to address the sectors not covered by the EU emissions trading scheme by, for example, the CO_2 tax and reimbursement in the case of taking appropriate measures such as more effective voluntary agreements and measures in the building sector.*

- *Curb energy demand growth in the transport sector by:*

  - *Decoupling transport activity growth from economic growth, considering a wide range of policies such as better urban planning, promoting teleworking, road pricing and modernisation of the economy away from transport-intensive activities.*

- *Fostering more energy-efficient modes such as railways.*
- *Encouraging car buyers to choose fuel-efficient cars and to retire old and inefficient cars by economic and regulatory measures, in particular vehicle taxation.*

Fossil Fuels

- *Continue to evaluate the competition situation in the oil market.*
- *On the basis of the newly established emergency legislation and the new Stockholding Agency (EGREP), make all necessary efforts, together with the oil industry, so that Portugal will be constantly compliant with the IEA emergency stockholding obligation in the future.*
- *Maintain the policy for diversified supply sources of natural gas and the balance between pipeline gas and LNG.*
- *Phase out subsidies and tax benefits for natural gas.*
- *Finalise the regulatory framework for the partial market liberalisation in July 2004. Set the schedule for the next steps of liberalisation.*
- *Make sure that third-party access to the facilities, including the pipelines, the LNG terminal and gas storage, is on a fair, non-discriminatory and transparent basis.*
- *Closely co-operate with Spain to introduce an Iberian gas market, preferably in pace with domestic market opening, and present a realistic time schedule for the market players.*

Renewables

- *Review the current feed-in tariff scheme in order to assure cost minimisation to consumers while ensuring investor confidence. Assess the benefits of incorporating incentives for cost reduction through gradually reducing the tariff level and the duration of the buy-back period. Also review the interaction of the feed-in tariffs and subsidies to determine when and which incentives can best be reduced.*
- *Assess progress towards a competitive renewable energy sector with a view to ensuring a stable investment environment until targets are met. Phase out the subsidies in the longer term when the different positive and negative externalities of renewables and other energy forms have been internalised.*
- *Continue efforts to streamline licensing procedures, including the environmental impact assessment, for renewable energy projects.*

- *Investigate the requirements of reliability and stability of the future electricity network, given the indicative goal of connecting large amounts of wind power to the grid.*

Electricity and Co-generation

- *Finalise the new legislation to carry out the objective of creating an Iberian market and to open the market to all consumers.*
- *Ensure the development of adequate interconnection capacity and its fair allocation.*
- *Consider at the Iberian level a mechanism to increase competition in generation through the use of "virtual power plant auctions" to make the energy available to suppliers other than major Iberian players.*
- *Monitor price developments in the Iberian electricity market, and be prepared and able to act promptly should concerns about manipulation of electricity prices arise.*
- *Develop a timetable for the phase-out of regulated power supply tariffs at the Iberian level and consistent with the 2003 EU Electricity Directive.*
- *Monitor generation investment developments in the Iberian electricity market. Consider removing the capacity payment or, as a temporary measure, replacing it with a more efficient instrument.*
- *Encourage the development of demand response mechanisms to enhance the security of supply and decrease the volatility of electricity prices.*
- *Re-evaluate the feasibility of the co-generation target to ensure that it is based on useful heat demand. Improve the environmental performance of some co-generation projects by encouraging the use of natural gas instead of oil.*

Research and Development

- *Develop a coherent energy R&D strategy with adequate financing to support energy policy objectives.*
- *Better monitor and assess the R&D projects.*
- *Improve the co-ordination between the different ministries involved in energy and related R&D.*
- *Improve the collection of data on governmental R&D funding.*

ORGANISATION OF THE REVIEW

REVIEW TEAM

The International Energy Agency (IEA) 2004 in-depth review of the energy policies of Portugal was undertaken by a team of energy policy specialists drawn from IEA member countries. The IEA review team visited Portugal in January 2004 for discussions with the Energy Administration, energy industries and non-governmental organisations.

The members of the team were:

Mr. Yasuo Tanabe
(Team Leader)
Director, International Affairs
Division, Agency of Natural
Resources
and Energy
Ministry of Economy,
Trade and Industry
Japan

Mr. Philippe Guillard
Head, Department of Refining
and Petroleum Logistic
Ministry of Economy,
Finance and Industry
France

Ms. Vera Pieterman
Project Manager, Energy Strategy
and Consumption Directorate
Ministry of Economic Affairs
The Netherlands

Mr. Lars Valdresbråten
Advisor, Oil Section
Ministry of Petroleum and Energy
Norway

Mr. Manfred Decker
Principal Administrator
Directorate-General for Energy
and Transport
European Commission

Mr. Peter Fraser
Senior Policy Advisor,
Energy Diversification
International Energy Agency

Mr. Jun Arima
Head, Country Studies Division
International Energy Agency

Ms. Lea Gynther
Desk Officer for Portugal
Administrator, Country Studies Division
International Energy Agency

Lea Gynther managed the review and drafted most of the report. Peter Fraser drafted the chapter on Electricity and Co-generation. Monica Petit and Bertrand Sadin prepared the figures and Marilyn Ferris provided editorial assistance.

ORGANISATIONS VISITED

The team held discussions with the following organisations:

- Associação Nacional de Concervação de Natureza (Quercus, an environmental NGO)
- Associação Portuguesa de Cogeneração (COGEN Portugal)
- Associação Portuguesa de Energias Renováveis (APREN, renewable energy association)
- Associação Portuguesa de Grandes Consumidores de Energia Eléctrica (APIGEE, large electricity consumers' association)
- Electricidade de Portugal (EDP):
  - EDP
  - EDP Energia
  - EDP Cogeneração
- Energy Services Regulatory Authority (ERSE, Entidade Reguladora dos Seviços Energéticos)
- Entidade Gestora de Reservas Estratégicas de Produtos Petroliferos (EGREP)
- GALP Energia
- GALP Energia ARL
- GALP Energia Transgás
- GALP Energia Gás de Portugal
- Grupo Estudos Ordenamento Territorio (Geota, an environmental NGO)
- Ministry of Economy
- Ministry for Public Works, Transport and Housing
- Ministry for Science and Higher Education
- Ministry for Cities, Spatial Planning and the Environment (MCOTA)
- Rede Eléctrica Nacional, S.A. (National Electricity Grid, REN)
- Sonae Comercialización de Energía (Sodesa)

- The National Institute of Industrial Engineering and Technology (INETI)
- The Portuguese Association of Consumer Protection (DECO)

The assistance and co-operation of all participants in the review are gratefully acknowledged.

REVIEW CRITERIA

The IEA *Shared Goals*, which were adopted by IEA Ministers at their 4 June 1993 meeting in Paris, provide the evaluation criteria for in-depth reviews conducted by the Agency. The IEA *Shared Goals* are set out in Annex B.

Figure 1

Map of Portugal

GENERAL ENERGY POLICY

OVERVIEW

Portugal is comprised of the mainland in the Iberian Peninsula and the Madeira and Azores archipelagos in the Atlantic Ocean. In 2002, the population of Portugal was 10.4 million, 5% over the 1990 level. In 1991-2001, however, the number of households increased by almost 19%. Half of the population lives on the coastal strip between Lisbon and Braga. The country's surface area is 92 000 km^2.

In 2002, GDP per capita in Portugal, measured using current purchasing power parities, was US$ 17 800[1] which is 71% of the OECD average. The Portuguese economy grew on average by 2.5% a year in 1990-2002. Fast growth in the second half of the 1990s (with GDP growing by 3.6% a year in 1996-2000) was followed by a sharp downturn led by weak domestic demand and, to a lesser extent, by the sluggish performance of the world economy. Activity contracted in 2003. As the population grew very little since 1990, GDP per capita increased by about one-third.

Portugal has a mild climate which reduces the need for heating; average annual air temperatures range from 7 °C for the central interior highlands to 18 °C on the southern coast.

ENERGY MARKET

In 2002, total primary energy supply (TPES) in Portugal was 26.4 Mtoe, up by 49% from the 1990 level. This is much more than the 35% GDP growth over the same period. Dependence on oil has declined from 75.4% in 1973 to 62.1% in 2002, which is still substantially higher than the IEA average of 41% (2001). Natural gas demand started to build up only after pipeline supplies from Algeria started in 1997 and, by 2002, gas gained a 10.3% share in TPES. The share of coal in TPES is 13%, down from 15.5% in 1990 and the share of combustible renewables and wastes 11%, down from 14% in 1990.

Domestic energy production was 3.6 Mtoe, accounting for 14% of TPES in 2002. The most important domestic energy source is combustible renewables and wastes (78%) followed by hydropower and small amounts of geothermal energy, wind and solar.

The reference scenario of the latest energy forecast includes the impacts of current policies and measures. Under these assumptions it projects an increase in energy demand of 14% between 2002 and 2010, with energy demand

1. On average in 2003, €1 = US$ 1.126.

Figure 2

Total Primary Energy Supply, 1973 to 2010

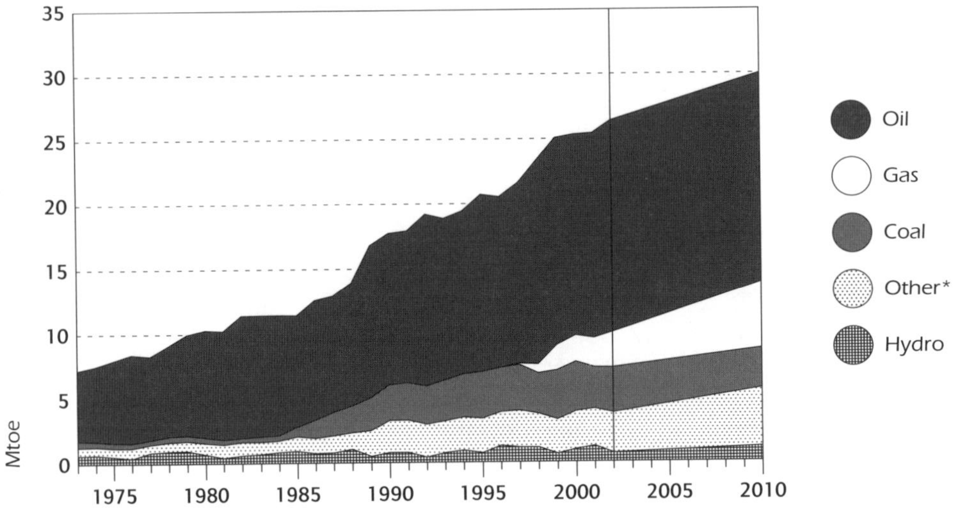

* includes geothermal, solar, wind, combustible renewables and wastes and electricity and heat trade.
Sources: *Energy Balances of OECD Countries*, IEA/OECD Paris, 2003; and country submission.

Figure 3

Energy Production by Source, 1973 to 2010

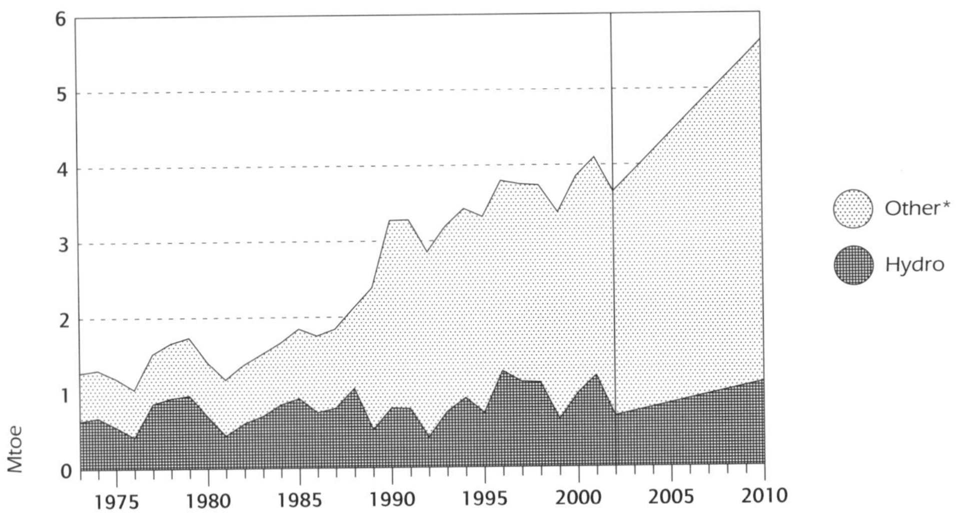

* includes geothermal, hydro, solar, wind, combustible renewables and wastes.
Sources: *Energy Balances of OECD Countries*, IEA/OECD Paris, 2003; and country submission.

growing much less than the economy (28%). The share of coal in TPES is expected to decline to 10% and oil to 54%, whereas gas is likely to attain 17%. Indigenous resources – all renewable energy – are expected to represent 19% of total supply. TPES is forecast to grow by a further 2.8 Mtoe by 2015 with gas increasing its share to 21%, renewables having a 16% share and oil demand remaining at around 53%.

The trends in total final energy consumption (TFC) and energy end-use efficiency are discussed in detail in Chapter 5.

ENERGY POLICY ADMINISTRATION

There have been many changes over the last four years in the energy policy administration and the institutional setting affecting the energy sector.

The General Directorate for Energy (DGE) within the Ministry of Economy was in charge of energy issues until it was replaced by a new body – General Directorate for Geology and Energy (DGGE) within the same ministry. DGGE has the competences of both the previous General Directorate for Energy and the former Institute for Mines and Geology. In the energy sector, DGGE is responsible for:

- Proposing legislation to regulate the energy sector and supervising its implementation.
- Granting licences and authorisations to energy-based plants and infrastructure and establishing their technical standards.

The Ministry for Public Works, Transport and Housing has replaced the former Ministry of Social Equipment. It is responsible for the policies in the housing and transport sectors, including the management of the infrastructures of the different modes of transport. The Ministry for Cities, Spatial Planning and the Environment has general responsibility for GHG emissions policy. The Ministry of Science and Higher Education has the responsibility for the national science, research, development and technology policy.

The regulation of the electricity and gas sectors has been entrusted to the Energy Services Regulatory Authority (*Entidade Reguladora dos Serviços Energéticos*, ERSE). It was initially established as an electricity sector regulator for the mainland in 1995 and started functioning in 1997. In 2002, its competences were extended to natural gas regulation. On this occasion, its authority was also extended to the Autonomous Regions of Madeira and the Azores. ERSE is a collegial body comprising a chairperson and two commissioners. They are appointed by government for a renewable five-year term and their mandate cannot be revoked. They are subject to an incompatibility regime that prevents them from having any economic interest in the industry. ERSE's decisions can be appealed before an administrative court only, so that the Minister of Economic Affairs cannot repeal its decisions.

ERSE decisions are made after listening to both the regulated parties and a consultative body representing various social interests. ERSE is financed through a surcharge on third-party access tariffs. In 2003, its budget was €6.4 million and it had a staff of 50. Despite the expansion of its responsibilities in 2002, additional staff has not been authorised up to now.

The Ministry of Economy is responsible for the formulation of the regulatory framework and licensing. ERSE is responsible for issuing codes for tariffs, commercial relations, network and interconnection assets and dispatch. ERSE is also responsible for setting the regulated prices of electricity and network services of electricity and gas.

The Competition Authority (*Autoridade da Concorrência*) was created in March 2002 to monitor and control compliance with competition legislation.

The Energy Conservation Centre, created in 1984, was transformed into the Agency for Energy (ADENE) by Decree-Law 223 of 2000. ADENE's main tasks are to prepare studies on energy efficiency and renewables, to promote efficiency labelling of electrical appliances and building certification, and to disseminate information on these topics.

ENERGY POLICY OBJECTIVES

The government recently issued its guidelines for energy policy that are based on three strategic goals – energy security, sustainable development and national competitiveness, through the Resolution of the Council of Ministers 63/2003 of 28 April 2003. These strategic goals led to the following energy policy objectives:

- Liberalisation of the energy markets.
- Reduction of energy intensity.
- Reduction of the energy bill.
- Improvement of the quality of service.
- Security of energy supply.
- Diversification of the types and sources of energy and use of domestic energy sources.
- Minimisation of the environmental impact.
- Improvement of national productivity.

Portugal faces the challenge of improving the competitiveness of its economy in the context of globalisation and enlargement of the European Union to new countries. These are common challenges to most IEA countries but, as a country benefiting from the EU Cohesion Fund, Portugal has enjoyed EU subsidies for the development and modernisation of its energy sector. One of the means of achieving better competitiveness is liberalisation of the energy market and its efficient regulation. Therefore, the government put great emphasis on creating the Iberian Electricity Market (see Chapter 8). Another is the final step in liberalising the oil market by removing the remaining price control on oil products (see Chapter 6).

Another challenge for Portugal is the climate change issue. Following its ratification of the Kyoto Protocol in 2002, the Climate Change National Programme has been proposed (see Chapter 4).

PROGRAMME OF INCENTIVES TO THE MODERNISATION OF THE ECONOMY (PRIME)

In 2000, a new Economic Operational Plan (OPE) was launched under the European Community Support Framework for the development of economic activities. The OPE contained two sub-programmes for the energy sector with a €1.6 billion budget for the period 2000-2006, namely the "Measure to Support Modernisation and Development of Electricity and Gas Infrastructures for Public Use" (budget €0.7 billion) and the "Measure to Support the Harnessing of the Energy Potential and Rationalisation of Consumption" (MAPE, budget €0.9 billion). MAPE's objectives are to promote energy efficiency, renewables and co-generation.

In July 2003, OPE was discontinued and replaced by the Programme of Incentives to the Modernisation of the Economy (PRIME). PRIME's total budget is €8.1 billion for the period from 2000 to 2006. OPE's energy sector activities were transferred to the new programme. PRIME's specific energy sector objectives are:

- Ensuring continuous energy supply at affordable prices.
- Reducing import dependence and fostering the development of local energy resources by using new and renewable energy sources.
- Reducing oil dependence and diversifying primary energy sources and supply origins, principally by promoting the use of other fuels such as natural gas.
- Fostering energy conservation and rational energy use and reducing energy intensity in all sectors, thus contributing to the reduction of the energy bill and import dependence.
- Reducing the environmental impact of energy production and use.
- Increasing the effectiveness and efficiency of energy products and services.

ENERGY SECURITY

Energy security is of great concern to Portugal's government because energy import dependence is very high. The government's main policies for enhancing energy security are:

- Reduction of energy import dependence through the promotion of renewables and energy efficiency.
- Diversification of energy sources and origins of imported energy.
- Guaranteeing adequate capacity of electricity supply.
- Oil stockholding.

One of the problems in terms of energy security is Portugal's frequent non-compliance with IEA's International Energy Program (IEP) obligations for more than a decade. This issue is discussed in detail in Chapter 6.

ENERGY TAXATION

Excise taxes are generally higher than the minimum levels set in EU legislation. In June 2002, the government increased the general VAT from 17% to 19%. This VAT rate applies to liquefied petroleum gas (LPG), gasoline and automotive diesel. Heating oil, electricity and natural gas for household use and diesel for agricultural purposes are subject to reduced VAT (see Table 1).

Table 1

Energy Taxes in Portugal, 2004[1]

Sector/fuel	*Excise taxes €/unit*	*VAT*[2] *%*
Households/electricity	0	5
Households/natural gas	0	5
Households/heating oil	90/1 000 litres	12
Household use/LPG	7.48/tonne	19
Industry/electricity	0	0
Industry/natural gas	0	0
Industry/low-sulphur fuel oil	15/tonne	0
Industry/steam coal	0	0
Industry use/LPG	0	0
Industry and commercial use/diesel	308.29/1 000 litres	19[3]
Non-commercial use/gasoline	522.6/1 000 litres	19
Non-commercial use/diesel	308.29/1 000 litres	19
Transport use/LPG	50.8/1 000 litres	19[3]

1 After the revision of 16 February 2004.

2 Because VAT is reimbursed for industry, it is shown as zero.

3 VAT is reimbursed 100% for heavy passenger vehicles, public transport, diesel oil for machines and agricultural tractors. For purposes other than private cars, it is reimbursed at a rate of 50%.

Sources: *Energy Prices and Taxes*, IEA/OECD Paris, 2003; and the Ministry of Economy.

CRITIQUE

Since the last in-depth review, there have been many positive developments in Portugal's energy sector. Natural gas, which was introduced in 1997, is successfully penetrating the energy mix through the extension of the pipeline networks and the construction of an LNG terminal and underground storage capacity. Important political decisions have been taken to create the Iberian electricity market (MIBEL). It has also been decided to fully liberalise the electricity market from July 2004. The regulator (ERSE) is functioning well with high independence and strong authority. A national action strategy to tackle climate change has been proposed.

However, Portugal is facing many challenges. For more than a decade, Portugal has been non-compliant with the IEP obligations. Energy intensity in terms of TPES per GDP has been growing. The share of oil is still significantly higher than the IEA average. Many preconditions need to be fulfilled for the effective functioning of the Iberian energy markets, fully liberalised domestic electricity markets and gas market reform. Simultaneous achievement of further economic development and meeting the Kyoto target is very challenging. These challenges are discussed in detail in the following chapters.

Portugal has very little indigenous energy sources with the exception of renewables. That is why the government has been making vigorous efforts to diversify the energy mix and supply sources. Increased use of natural gas, including the recent introduction of LNG, and the promotion of renewables have been contributing positively to energy security as well as to climate change mitigation and environmental protection. However, noting the continuing high dependence on imported oil, further diversification should be explored. While it should be mainly up to the market to decide the energy mix and supply under the liberalised energy economy, the government needs to monitor developments and provide an appropriate market framework to ensure energy diversification as well as environmental protection, including climate change mitigation. The introduction of an emissions trading system and carbon taxation, which is under consideration, would contribute to these ends.

One cross-cutting challenge is the enhanced communication among various stakeholders. A large number of ministries, public institutions, associations and enterprises are involved in energy issues. Efficiency in decision-making and policy implementation should be increased and energy should be better taken into account in other related policies. Closer co-ordination between ministries is essential, in particular in the transport sector and R&D. Consumers also need to be fully involved in the policy-making process and relevant information must be disseminated. This is particularly important for the success of market liberalisation and climate change mitigation where they have a major role to play.

The decision to create MIBEL would be a very important step in creating a competitive market and, therefore, is highly commendable. However, many important issues need to be solved before it becomes fully functional (see Chapter 8).

The creation of MIBEL will increase the necessity for a consistent approach to electricity and natural gas policies between Portugal and Spain. Since one of the objectives of market enlargement is to rationalise investment, a common approach to the security of supply policy is particularly important. A consistent approach is also essential in climate change mitigation from the viewpoint of competitiveness, noting the close economic integration between the two countries.

One major initiative of the government is the planned merger of the natural gas activities of GALP into EDP. Through this, EDP will have access to both gas customers and gas supply contracts of Transgas (the supply and transmission company). While such a structure appears logical from a commercial point of view in the Iberian market, its implication on competition in the domestic electricity and gas markets needs to be scrutinised. A combination of structural separation and regulatory oversight to ensure non-discriminatory access to natural gas supplies for power generators will be important. Separation of the gas transmission system operator from the owner of the gas supply contracts will be an important step.

Given ERSE's new responsibilities in gas market regulation and many challenging tasks in electricity market reform, it is important that ERSE is given the necessary means to conduct its duties also in the gas sector.

RECOMMENDATIONS

The government of Portugal should:

- *Continue efforts to diversify the energy mix and closely monitor progress to improve security of supply and enhance climate change mitigation and environmental protection.*
- *Enhance co-ordination of energy policy measures between the different ministries and appropriate organisations to take better account of energy in other policies and to increase efficiency.*
- *Involve all stakeholders, including consumers, in developing energy policies and disseminate information to them.*
- *Continue the close co-operation with the government of Spain to develop consistent policies for the Iberian energy markets. This includes electricity*

tariff schemes, the access and availability of natural gas, the allocation of CO_2 emission permits, and the policies regarding renewables and security of supply.

- *Scrutinise the impacts of EDP's acquisition of GALP's gas supply and retailing activities on competition in the electricity and natural gas markets.*
- *Make sure that ERSE is given the necessary means to conduct its duties, also in the gas sector.*

ENERGY AND THE ENVIRONMENT

CLIMATE CHANGE

GREENHOUSE GAS EMISSIONS

In 2001, Portugal's total GHG emissions[2] were 83.8 million tonnes of CO_2-equivalent. which is 36.4% above the 1990 level (61.9 Mt CO_2-eq.). Its energy-related CO_2 emissions increased by 23.4 Mt[3] (49%) since 1990, reaching 63 Mt in 2002. This was the main driver of the GHG increase over the period. CO_2 emissions accounted for about 77% of the total GHG emissions in 2001.

When broken down by fuel, 68.5% of CO_2 emissions stemmed from the use of oil, 21.4% from coal and 10% from natural gas in 2002 (see Figure 4). In an analysis by sector, public electricity and heat production accounted for 35% of energy-related CO_2 emissions, followed by transport 30%, industry 19%, "other" sectors (services and agriculture) 9%, the residential sector 3.5% and other energy industries with a 3.5% share (see Figure 5). CO_2 emissions from the "other" sectors grew the fastest in 1990-2002 by 126%, but a rapid increase could be observed in all sectors: transport (91%), public electricity and heat production (56%), other energy industries (47%), the residential sector (33%) and manufacturing industries and construction (20%).

In 2001, Portugal's energy-related CO_2 emissions per GDP were 0.35 kg of CO_2/US$ (using 1995 prices and purchasing power parities) – an increase of 9% since 1990 – whereas the IEA Europe average was 0.39 kg. CO_2 emissions per capita increased by 47% between 1990 and 2001, reaching 5.87 tonnes, which is still below the IEA Europe average of 7.6 tonnes.

CO-ORDINATION OF CLIMATE CHANGE POLICY

The Institute for the Environment within the Ministry for Cities, Spatial Planning and the Environment (MCOTA) is responsible for co-ordinating the national policy for climate change mitigation. The Climate Change Committee (CAC), an interministerial body, was created in 1998 to analyse the integration of climate change issues into sectoral policies. It published the National Strategy for Climate Change in May 2001 and, thereafter, it has been preparing the Climate Change National Programme (PNAC).

2. Carbon dioxide (CO_2), methane (CH_4), nitrous oxide (N_2O), perfluorocarbons (PFCs), hydrofluorocarbons (HFCs) and sulphur hexafluoride (SF_6).
3. This statistic is based on the Sectoral Approach.

Figure 4

CO_2 Emissions by Fuel*, 1973 to 2002

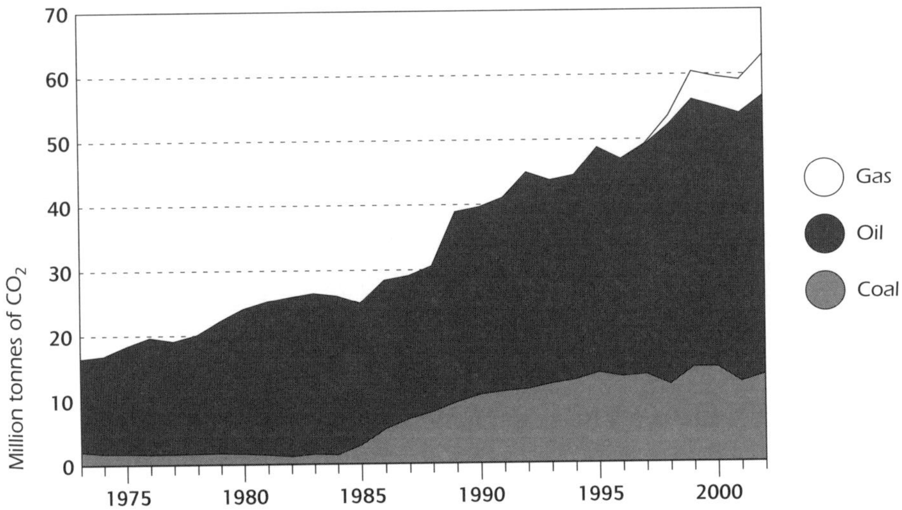

* estimated using the IPCC Sectoral Approach.
Source: *CO_2 Emissions from Fuel Combustion*, IEA/OECD Paris, 2003.

Figure 5

CO_2 Emissions by Sector*, 1973 to 2002

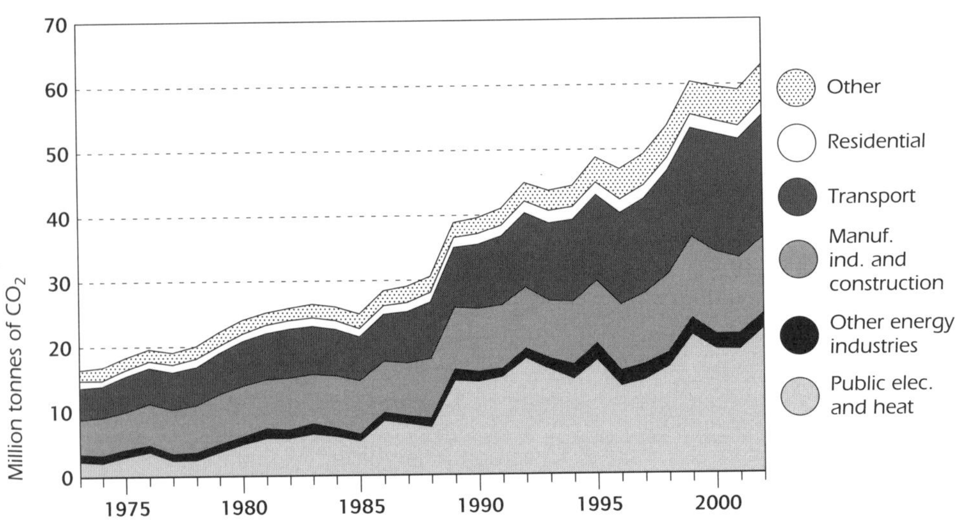

* estimated using the IPCC Sectoral Approach.
Source: *CO_2 Emissions from Fuel Combustion*, IEA/OECD Paris, 2003.

Figure 6

Energy-related CO_2 Emissions per GDP in Portugal and in Other Selected IEA Countries, 1973 to 2010

(CO_2 emissions/GDP using 1995 prices and purchasing power parities)

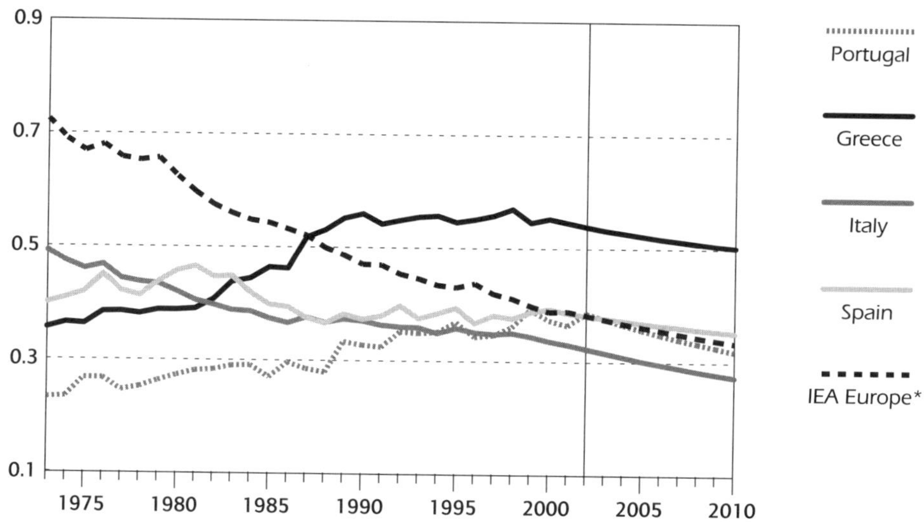

* excluding Norway from 2002 to 2010.

Sources: *Energy Balances of OECD Countries*, IEA/OECD Paris, 2003; *National Accounts of OECD Countries*, OECD Paris, 2003; and country submissions.

The National Observatory for Climate Change in Portugal was established by Law 93/2001 in August 2001 to compile, analyse and disseminate information, studies and research on the impacts of climate change. However, the Observatory has not yet become operational owing to budgetary and staffing issues.

Local authorities have an important role in implementing policies and measures. They also promote awareness on climate change and try to induce behavioural changes in individuals and in small and medium-sized enterprises.

CLIMATE CHANGE POLICIES

Portugal ratified the Kyoto Protocol in May 2002. Portugal's commitment under the EU Burden-Sharing Agreement under the Kyoto Protocol is to limit the increase of its GHG emissions to 27% between 1990 and the first commitment period of 2008-2012. Emission baseline studies for PNAC indicate that total GHG emissions will increase by 41 to 48% between 1990 and 2010, including the positive impact of existing measures of 14-16 percentage points. The principal policies in the baseline scenario are improving energy efficiency (see Chapter 5) and promoting the use of renewable energies (see Chapter 7). This means that to meet its target, Portugal needs to achieve an additional reduction of 14 to 21 percentage points, *i.e.* 8.6 to 12.6 Mt CO_2-eq. from the baseline scenario.

The PNAC incorporates existing and additional policies, measures and associated instruments which the government will use to meet its Kyoto target. Its adoption and implementation is, however, subject to the finalisation of the National Allocation Plan (NAP) for the EU Emissions Trading Scheme (EU-ETS) and Parliament's decision on the use of financial and fiscal instruments, as well as finalisation of the public consultation. The government expects the PNAC to be adopted in the summer of 2004. The guiding principles in the preparation of the PNAC have been the coverage of all the sectors, environmental protection, economic efficiency, competitiveness of enterprises, employment, and political and administrative feasibility.

PNAC consists of two blocks of policies and measures. Those in the "Immediate Block" are in the implementation or planning phase and will be in place by 2005. Measures which have already been implemented or are being implemented include, for example, the Energy Efficiency and Endogenous Energy (E4) programme, energy labels and standards for lighting and appliances, building regulations, vehicle taxation reform, modernisation of the inter-urban infrastructures of rail transport and promotion of public transport and modal shift from road transport (see Chapter 5). These measures are expected to reduce GHG emissions by 8.15 Mt CO_2-eq as compared to the baseline scenario. Measures in the "Additional Block" are completely new and will mainly cover activities that are not covered by the EU-ETS. Emissions reductions from the Additional Block are divided between the sectors in the following way: 42.9% for transport, 31.9% for agriculture and livestock, 23.3% for energy supply (non-trading activities) and 1.8% for forestry. No reductions are allocated to the residential and services sectors. The Additional Block, together with Kyoto flexible mechanisms, is planned to reduce GHG emissions by a further 7.4 Mt CO_2-eq. No separate emissions reduction estimates are presented for a CO_2 tax or the EU-ETS, but they will be part of the Additional Block.

There has been no comprehensive analysis of the cost-effectiveness of the individual policies and measures or blocks of measures, but some analysis has been conducted (see Table 2).

The EU-ETS directive was published in September 2003 but it has not yet been transposed to Portuguese legislation. No specific implementing agency for emissions trading has yet been established and the criteria and rules for participation have not yet been defined. The draft NAP, prepared by an inter-governmental working group, was issued for public consultation on 17 March 2004. It allocates a total of 116.7 Mt of CO_2 emission allowances

Table 2

Examples of Cost-effectiveness of Climate Change Mitigation Policies

Policy	*Emissions reduction in 2010[1], Mt*	*Cost estimate, million euros*	*Cost-effectiveness, euro/tonne of CO_2*
E4 Programme: energy efficiency measures	0.5-0.6		n.a.
E4 Programme: building 3 250 MW of new wind capacity and meeting the indicative targets for other renewables	3.3-3.7		44
E4 Programme: building further 500 MW of new wind capacity	0.3-0.9		from "win-win" to <35
Improving efficiency of electricity generation	0.7		"win-win"
Improving efficiency in electricity transmission and distribution	0.14-0.16	10 for transmission, 100 for distribution	from "win-win" to <15
Promoting CHP	0.19-0.25		n.a.
Improving efficiency of electricity consumption (reduction of 1.3 TWh)	0.46-0.56		from "win-win" to <15

[1] Emissions reduction due to the policy in 2010 from the level in the baseline scenario as compared to the 1990 level.

Source: Ministry for Cities, Spatial Planning and the Environment.

over the 2005-2007 period (38.9 Mt per year), and also provides for a 5.6 Mt (1.87 Mt per year) reserve for new entrants, which it will allocate on a first come, first served basis. Power generators receive 21.25 Mt CO_2 of emission allowances per year, which requires them to reduce emissions by 2.9% from 2002 levels. Unused allowances at the end of the period will be auctioned off.

By March 2004, 224 installations eligible to EU-ETS and representing 43% of national GHG emissions had been identified. Seventeen of them are power stations, two are refineries and 55 are co-generators representing 75% of the totality of the emission allowances to be awarded and 29% of total GHG emissions in 2002.

The National Strategy for Climate Change recognises that using joint implementation (JI) and clean development mechanisms (CDM) to some extent will be necessary. The strategy refers to the possibility of implementing CDM in Portuguese-speaking Africa. In the preparation of the PNAC, the possible contribution of JI and CDM has been estimated at between 1.7 and 5.6 Mt CO_2-eq. The government has appointed the NAP working group to facilitate the use of credits from these mechanisms and to prepare instructions to help the industry to participate. However, practical work has not yet started.

AIR POLLUTION

Sulphur oxides (SO_x), nitrogen oxides (NO_x) and non-methane volatile organic compound (NMVOC) emissions all increased in the 1990s. However, progress has been made in the ambient levels of air pollutants in urban areas, for example in Lisbon and Porto. Ground-level ozone, although not well monitored, seems to occur seasonally in most regions. Portugal now faces the challenge of meeting its international commitments (Gothenburg Protocol) to reduce emissions of SO_x by 54%, NO_x by 13% and NMVOCs by 22% by 2010 compared to 2000.

In December 2002, Portugal issued the Programme for National Emission Ceilings (PTEN) to define the strategy for implementing the EU Directive 2001/81/EC on emission ceilings for SO_2, NO_x, NH_3 and VOCs. There are several instruments in place in the PTEN with impacts on the energy sector, all following from the national implementation of respective EU directives. They are:

- Reduction of the sulphur content of fuel oil (EU Directive 93/12/EC). The law puts an upper limit of 1% on the sulphur content of fuels.
- Reduction of emissions from large combustion plants (EU Directive 2001/80/EC). As a consequence, desulphurisation plants will be installed at the Sines (EDP) and Pego (Tejo Energia) coal-fired power plants by 2007.
- Integrated pollution prevention and control (EU Directive 96/61/EC).
- Promotion of electricity generation from renewable energy in the electricity market (EU Directive 2001/77/EC).

ALTERNATIVE TRANSPORT FUELS

The guidelines for energy policy established in the Resolution of the Council of Ministers 63/2003 include the objective to develop energy-efficient and cleaner transport systems. Some action has been taken to promote the use of alternative transport fuels, namely natural gas, LPG, electricity and biofuels.

A series of pilot projects have been carried out mainly by municipalities as well as public and municipal enterprises aiming at the introduction of vehicles using alternative fuels in their fleets. The use of natural gas vehicles began in Braga and Porto in 2000 and in Lisbon in 2001. At present, there are over 30 000 vehicles using LPG, a few dozens using either natural gas (in the public transport fleets in Lisbon, Porto and Braga) or electricity, and only a few using biofuels, mainly in some pilot projects.

A vehicle tax exemption of 40% was introduced in 2000 for vehicles using exclusively LPG, natural gas or renewable energy. In 2001, the level of the exemption was increased to 50% for vehicles using exclusively LPG or natural gas. For vehicles with hybrid engines that use conventional fuels and LPG, natural gas, electricity or solar energy, a 40% reduction of the vehicle tax is provided. Lower taxes on LPG and natural gas, as compared to gasoline and diesel for road vehicles, provide an incentive to use these fuels (see Table 1 in Chapter 3), but there are no tax exemptions for biofuels.

The MAPE programme provides subsidies for the renewal of the public transport fleet with less polluting vehicles. The programme pays a maximum of half the additional cost of purchasing new gas or electricity-driven vehicles as compared to the cost of conventional vehicles. It is envisaged to extend the support to biofuel-driven vehicles whenever the mixing of biofuel in conventional fuel necessitates modifications to the engine and feeding system.

The EU Directive 2003/30/EC on the use of biofuels in the transport sector has not yet been transposed to the Portuguese legislation. The directive requires that biofuels should acquire a 5.75% share in the total sales of petrol and diesel for transport use by 2010. According to anecdotal evidence, about half of the necessary biodiesel could be produced from domestic crops.

CRITIQUE

Since the last in-depth review, there have been several developments in climate change mitigation policies. Portugal ratified the Kyoto Protocol, the National Strategy for Climate Change has been established and the Climate Change National Programme (PNAC) is advancing.

However, several factors indicate that meeting the Kyoto target will be a very challenging task for Portugal. First, in 2001, total GHG emissions were already 9.4 percentage points above the target level for 2008-2012. Second, even with the policies and measures in place (reference scenario), emissions are projected to grow to a level which exceeds the target by 14-21 percentage points requiring the prompt introduction of new strong policies and measures. Third, the baseline estimates should be reviewed when necessary to reflect the rapid growth of energy demand in some sectors. In the reference scenario, the increase in energy use from 2001 to 2010 is estimated only at 4.5 Mtoe, whereas the increase from 1990 to 2001 was 7.7 Mtoe. While GHG emissions increased by 21.9 $MtCO_2$-eq. from 1990 to 2001, the reference scenario forecasts only a 3.5-7.8 $MtCO_2$-eq. increase in GHG emissions by 2010. The lower expectations arise from more moderate projections for economic growth. However, in reality, energy use in Portugal has been growing faster than the economy. It is not clear that the trend can be reversed with the package of existing and planned policies.

The introduction of PNAC should be finalised without any further delay. It will be particularly important that PNAC includes very effective policies and measures with a clear implementation plan which should establish both time schedules as well as adequate funds for implementation. Conscientious monitoring of the trends and the effectiveness of the policies, followed by prompt policy response, is absolutely essential.

In the draft PNAC, the expected GHG emissions reduction from each policy is shown and cost-effectiveness analysis has been conducted for some policies and measures. However, it is not evident to what extent the results are used for minimising the cost of climate change mitigation. For example, the very generous support to renewable energy is not necessarily cheaper than energy efficiency policies. The targets of some policies and measures seem to be challenging, in particular the 18% target for CHP. While the draft PNAC is very ambitious and actively promotes measures in some sectors, it refrains from presenting strong measures for the transport sector. Furthermore, the cost-effectiveness of transport sector policies has not been evaluated even though the reductions from this sector are expected to bring almost half the necessary reduction in the Additional Block.

Introduction of the EU-ETS will be a good incentive to curb GHG emissions from the large emitters in the energy and industrial sectors (the so-called trading sectors). A draft National Allocation Plan (NAP) has been issued for public consultation. It allows for a 1.6% annual increase in emissions for the eligible companies as a whole. At the same time, the business-as-usual scenario assumes an approximately similar annual increase in emissions. Effectively, the NAP puts less pressure on the trading than on the non-trading sector. For this reason, and because the EU-ETS covers less than half of the total GHG emissions, additional emissions reduction measures in the non-trading sectors are necessary (see Chapter 5).

Portugal envisages the use of the Kyoto flexible mechanisms to complement domestic efforts. This is a sensible policy as the use of these mechanisms is likely to reduce the cost of achieving the target. Their use is also necessary because the reference scenario shows that domestic measures are not enough to meet the target. It is even possible that Portugal has to eventually rely even more heavily on these mechanisms than envisaged if further study of the emission trends shows faster-than-expected growth. However, the necessary preparatory work has not yet started. Given the rather complicated procedure of using the mechanisms, a late start could make their use in time for 2008-2012 difficult. Some countries, such as the Netherlands, have already established the necessary procedures and identified eligible projects which provide a learning opportunity.

Current taxation (see Table 1) does not fully address environmental externalities related to climate change and air pollution, which is still a problem in Portugal. The government is considering the introduction of a CO_2 tax to address emissions from the sectors which will not participate in emissions trading. Given the level of challenge the country is facing, this appears to be a sensible approach and the discussion should be accelerated. Such a tax may be introduced by making it revenue-neutral.

Excise taxes on fuels also deserve attention as they are highly favourable to diesel. In the existing car fleet, diesel engines emit more harmful pollutants such as particles and NMVOCs with health risks even at fairly low concentrations. In this context, it should be reviewed whether the taxation favouring diesel is sustainable but giving due consideration for the availability of new, cleaner diesel technologies and tightening emission standards which will, over time, reduce diesel's externalities. In doing so, the impact of any tax revision on the demand for diesel and gasoline should be assessed. If it is likely to result in more GHG emissions, then other measures should be introduced to offset such an impact. To reduce the negative externalities of diesel, the vehicle taxation could be made to favour diesel vehicles with cleaner technologies.

The EU directive on the use of biofuels remains to be transposed to the national legislation. It will also be necessary to evaluate the effectiveness and cost of the policies in place for meeting the target level. In this context, the taxation of biofuels could be made more in line with other fuels with low CO_2 emissions.

RECOMMENDATIONS

The government of Portugal should:

- *Finalise PNAC and adopt a time schedule for the policies and measures with reasonable and achievable goals at each step. Periodically review it on the basis of recent GHG projections and the progress of various policies and measures.*
- *Further examine the cost-effectiveness of climate change policies and measures.*
- *Clarify the role which Kyoto flexible mechanisms should have in meeting the emissions targets and accelerate the preparation for their utilisation.*
- *Increase the use of fiscal instruments to internalise the externalities related to climate change and air pollution.*

ENERGY DEMAND AND END-USE EFFICIENCY

DEMAND TRENDS

Portugal's energy intensity (TPES per unit of GDP) has increased since 1990, particularly towards the end of the period (2002), but still remains below the IEA Europe average (see Figure 7).

Figure 7

Energy Intensity in Portugal and in Other Selected IEA Countries, 1973 to 2010

(toe per thousand US$ at 1995 prices and purchasing power parities)

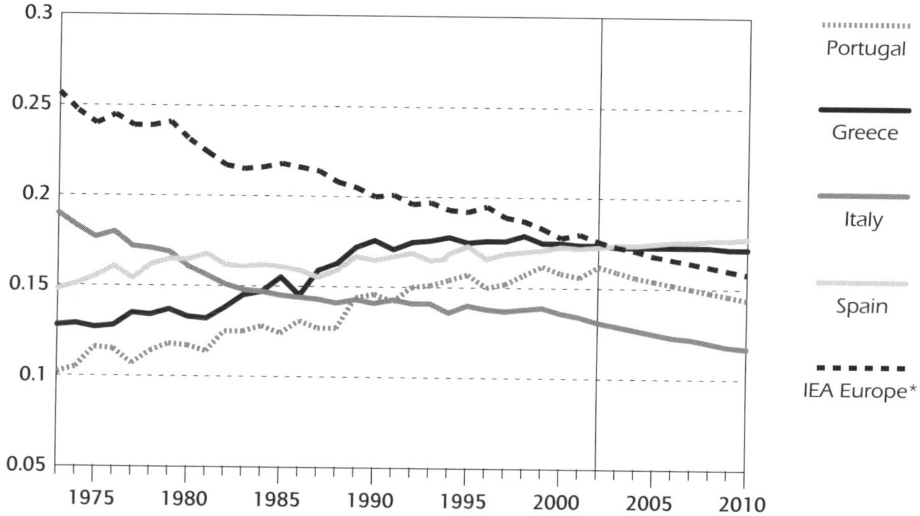

* excluding Norway from 2002 to 2010.

Sources: *Energy Balances of OECD Countries*, IEA/OECD Paris, 2003; *National Accounts of OECD Countries*, OECD Paris, 2003; and country submissions.

Total final consumption of energy (TFC) was 20.8 Mtoe in 2002, up by 48% from the 1990 level. In 2002, oil accounted for 63.4% of TFC, electricity 17.2%, combustible renewables and wastes 11.9%, natural gas 5.7%, heat 1%, coal 0.9% and other renewables 0.1%, as shown in Figure 8. Between 1990 and 2002, the share of combustible renewables and wastes in TFC declined from 16.6% and coal from 4.2%, while the share of gas increased from 0.4% (town gas in 1990) and electricity from 14.5%. The share of oil remained almost level.

The industry (including non-energy use of 0.75 Mtoe) is the largest energy-consuming sector (41%), followed by transport (33%), the residential sector (14%), the services sector (9%) and the agricultural sector (2%) (see Figure 9).

Industrial energy consumption (including non-energy use) increased by 38% between 1990 and 2002 reaching 8.5 Mtoe. Growth in industrial production over the same period was 23%. The government estimates that energy consumption in this sector will increase by 11% from 2002 to 2010. Two-thirds of industrial energy is used in the production of cement, iron and steel, nitrogen fertiliser and paper and board.

Consumption in the transport sector grew dramatically by 80% between 1990 and 2002, reaching 6.9 Mtoe. For the period from 2002 to 2010, the government expects only a 20% increase. Both passenger and freight transport volumes have increased significantly since 1990 and growth is expected to continue (see Table 3). Growth of consumption in the transport sector has been consistently greater than the EU average, and greater than in Greece and Spain which are also EU cohesion countries. According to the European commission, in 2001 the number of passenger cars in Portugal was 364 per 1 000 inhabitants, whereas the EU average was 488. The vehicle stock increased from 1.85 million in 1990 to 3.75 million in 2001. However, the Portuguese vehicle statistics are somewhat uncertain as they are collected through insurance information. Most freight transport is by road.

Table 3

Trends in Public and Private Transport, 1990 to 2010

	1990	*1995*	*2000*	*2010*
		Passenger transport, billion passenger-km		
Cars	36.6	56.3	76.2	111.1
Buses	13.3	10.7	12.0	13.9
Rail[1]	7.3	6.0	5.6	8.5
Total	**57.2**	**73.0**	**93.8**	**133.5**
		Freight transport, billion tonne-km		
Road	13.5	16.9	25.0	33.8
Rail	1.2	1.5	1.9	3.2
Total	**14.7**	**18.4**	**26.9**	**37.0**

[1] Includes trains, subway and trams.

Source: Ministry of Public Works, Transport and Housing and the Ministry of Economy.

Figure 8

Total Final Consumption by Source, 1973 to 2010

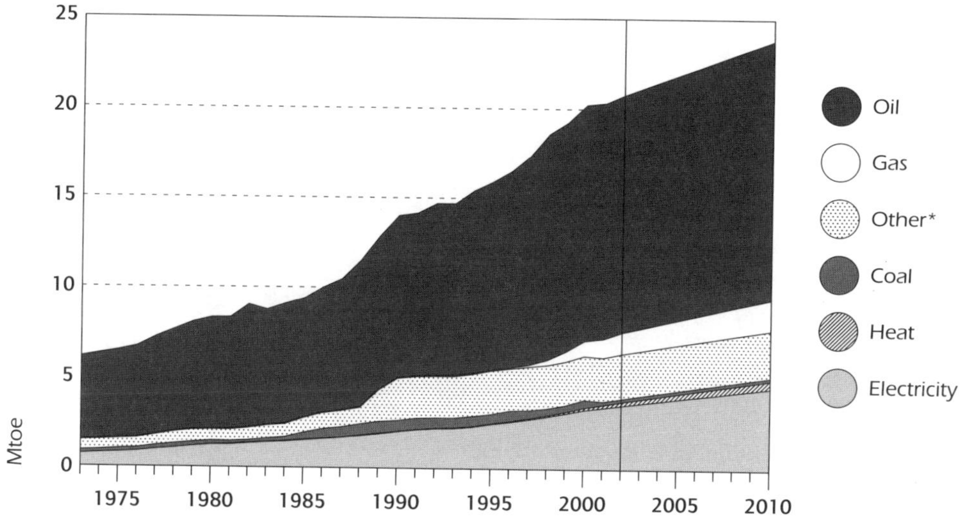

* includes geothermal, solar, wind, combustible renewables and wastes.

Sources: *Energy Balances of OECD Countries*, IEA/OECD Paris, 2003; and country submission.

Figure 9

Total Final Consumption by Sector, 1973 to 2010

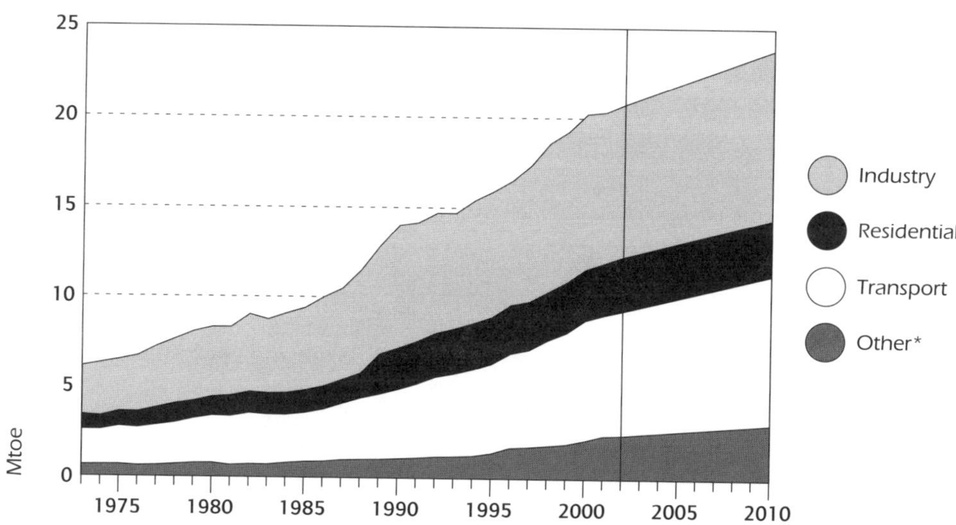

* includes commercial, public service and agricultural sectors.

Sources: *Energy Balances of OECD Countries*, IEA/OECD Paris, 2003; and country submission.

Figure 10

Total Final Consumption by Sector and by Source, 1973 to 2010

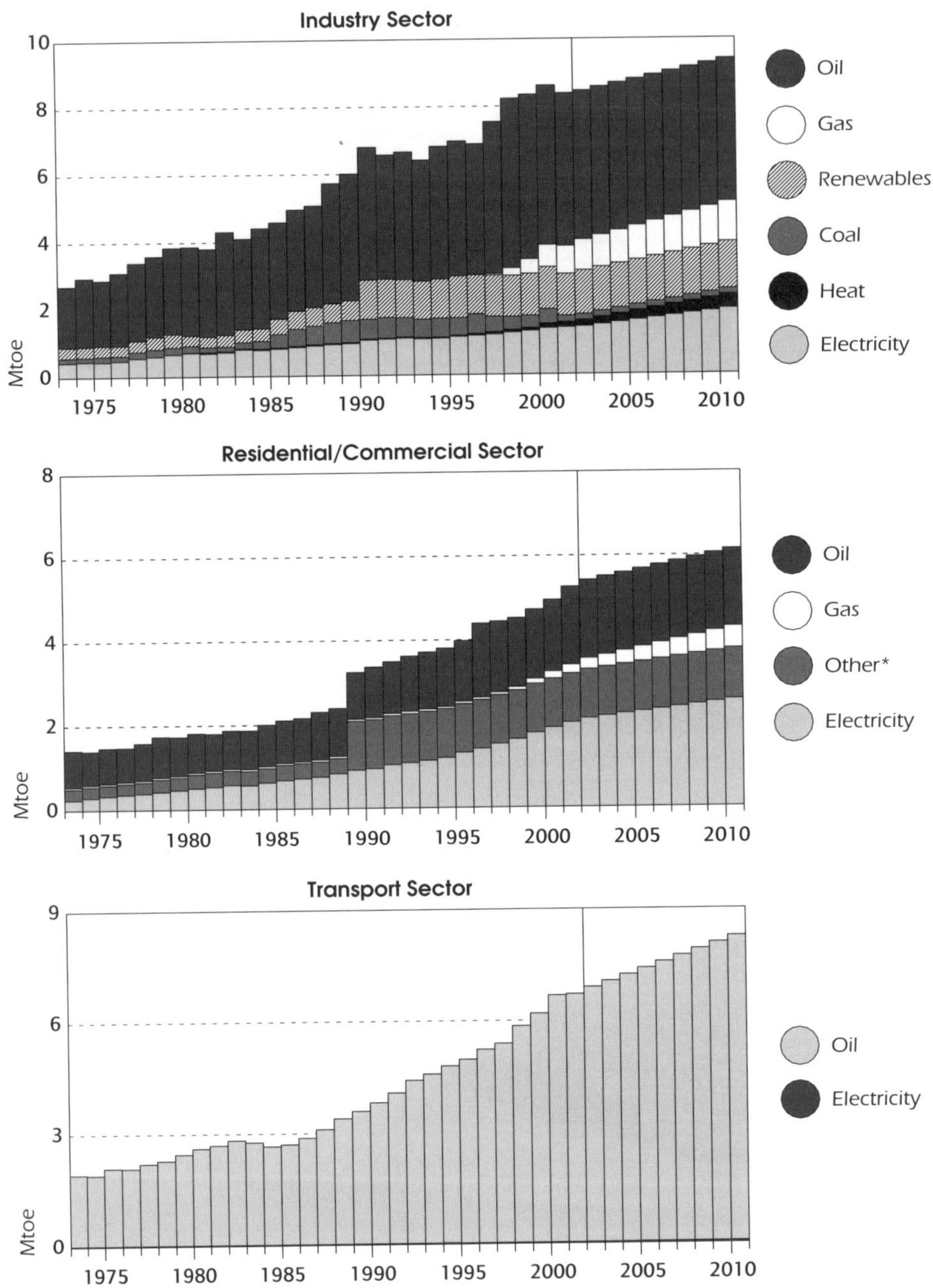

* includes geothermal , solar, wind, combustible renewables and wastes.

Sources: *Energy Balances of OECD Countries*, IEA/OECD Paris, 2003; and country submission.

Between 1990 and 2002, energy consumption in the residential sector increased by 31%, reaching 3 Mtoe. However, electricity demand in this sector grew by 92% over the same period. There are several factors which caused this large increase. The number of dwellings increased by about 20%, and at the same time, households are using more energy, particularly electricity, with a larger number of appliances and increasing use of air-conditioning. In 2000, it was estimated that 40% of service sector facilities and 5% of households had air-conditioning. Air-conditioning use has been increasing steadily at more than 500 000 new units installed every year and the trend is expected to continue for several years to come. Because of the mild climate, little energy (approximately 0.8 Mtoe in 2000) is used for space heating. Buildings are heated with wood, electricity and liquefied petroleum gas, but heating oil was recently introduced in the market to provide more consumer choice. From 1990 to 2002, energy consumption climbed by 221% in the services sector, reaching 1.9 Mtoe. Energy use in the agricultural sector was 0.48 Mtoe, up by 3% from the 1990 level. The government expects only a 14% increase in total demand in the residential, services and agricultural sectors from 2002 to 2010.

ENERGY EFFICIENCY POLICY

CROSS-CUTTING ACTIVITIES

One of the key programmes is the Energy Efficiency and Endogenous Energy Programme (E4 Programme) which was introduced in October 2001. While it also supports renewable energies, it has the following energy efficiency activities:

- Financial incentives to projects on energy efficiency, energy infrastructure and R&D on energy.
- National Programme for Energy Efficiency in Buildings.
- Development of more efficient and cleaner transport systems.
- Information dissemination to consumers.
- Issuing the legal framework for micro-generation.
- Facilitating the connection of decentralised generation, including co-generation, by adapting the investment programmes of grid operators.

Financial incentives for energy efficiency projects, co-generation and fuel conversion projects are given by the Measure to Support the Harnessing of the Energy Potential and Rationalisation of Consumption (MAPE). It runs from October 2001 to 2006 within the framework of the Programme of Incentives to the Modernisation of the Economy (PRIME) (see Chapter 3). MAPE replaced the former scheme called the Incentive System for the Rational Use of Energy

(SIURE). It provides both grants and zero-interest loans. Additional grants and interest-free loans for energy efficiency projects can be acquired from the System of Incentives for Modernising Enterprises (SIME) if the project is part of a wider industrial modernisation package.

The government has issued an action plan to foster the implementation of energy policies, including energy efficiency, at the local level. To do this, local agencies were established after October 2001 under the EU Save Programme. Financial support to energy efficiency projects of the municipalities was provided from MAPE. During the last five years, 88 projects received €12 million of support and led to investments of €29 million. The energy savings are estimated at 9.5 ktoe per year.

Energy efficiency is one of the main areas of Portugal's energy R&D (see Chapter 9). It receives one-third of the non-nuclear energy R&D budget.

Table 4

Subsidies for Energy Efficiency Projects, 2001-2003

Programme	*Number of projects*	*Subsidies (million euros)*	*Private Investment (million euros)*	*Energy savings (ktoe/year)*	*CO_2 savings (kton/year)*	*Subsidies (€/t CO_2)*
			MAPE			
Co-generation	19	9.3	75.5	148	348	3
Energy efficiency (industry)	19	1.8	5.7	15	42	50
Energy efficiency (services)	11	1.6	8.7	1.8	10.5	150
			SIME			
Energy efficiency	668	11.4	57.2	..	..	

Source: The Ministry of Economy.

INDUSTRY

The main policy instrument in the industrial sector is the Management Regulation for Energy Consumption (RGCE) which has been in force since 1982. It requires energy-intensive industrial companies to progressively reduce their specific energy consumption. The main industrial sub-sectors covered are textiles, ceramics and glass, food and beverages, chemicals and cement. RGCE requires these industries to audit the plants' energy consumption twice a decade, to prepare and carry out annual plans for rationalising energy consumption and to meet their targets on energy savings if they wish to receive subsidies for energy efficiency investments. In 2005, the above-mentioned industries will be subject to RGCE's obligations regardless of their intention to apply for subsidies.

In 2001-2003, 191 audit reports were submitted to the Ministry of Economy, in addition to the 549 already conducted by the end of 2000. The audits show that energy savings of 37 ktoe per year have been achieved corresponding to 3% of the energy consumption of the audited plants. The measures used in the industries include energy management systems, co-generation projects, thermal insulation, heat recovery systems and the installation of more efficient equipment.

Co-generation plants reaching certain efficiency levels are promoted by favourable feed-in tariffs. This is discussed in more detail in Chapter 8.

There are very few measures in place for small and medium-sized industries. Those implemented are information dissemination and energy auditing which has been conducted in some small companies in the textiles, ceramics, dairy, wood and cork industries.

TRANSPORT

Policies in the transport sector are the following:

- **The Management Regulation for Energy Consumption:** The RGCE has also been implemented in public and private transport enterprises where energy consumption exceeds 500 toe per year. The RGCE requires these companies to carry out energy audits and to publish a plan for the rationalisation of their energy use every three years.
- **Car labelling:** The EU Directive 1999/94/EC on consumer information on fuel economy and CO_2 emissions was transposed to national legislation by Decree-Law 304/2001. Consumers are now able to compare the energy efficiency and CO_2 emissions when buying new vehicles.
- **Investments in improving public rail transport and promotion of modal shift:** In the last few years there has been increased investment in public passenger transport which is generally not yet considered to provide a good level of service. In Porto, an underground system started operation in 2003 and extensions are under way. The system in Lisbon is being extended at present and further extensions are planned. An agreement with Spain has been signed to develop a high-speed railway link between the two countries and public tendering is scheduled for 2005. Other activities in the railway system include electrification, renewal of the networks and the fleet, and construction of railway access to ports, airports and large industrial facilities.

- **Mandatory vehicle inspections:** Control of vehicle energy efficiency during the mandatory periodic inspections has been implemented.
- **Eco-driving:** Portugal tests the teaching of eco-driving in driving schools.
- **Car fleet renewal:** Scrapping fees are provided and change of taxation to promote the renewal of the car fleet is at present being considered.
- **Strengthening transport administration:** Lisbon and Porto have established new local authorities to manage transport issues. In addition to contracting public transport operators, they work to reduce the use of private cars. Transport policies, however, continue to be under two different ministries, namely the Ministry for Public Works, Transport and Housing (public transport) and the Ministry of Internal Administration (private transport).
- **Road pricing:** The local authority in Lisbon is considering introducing road pricing in historical areas.
- **Avoiding "empty" freight transport:** The objective is to reduce the number of empty return trips of private trucks by, for example, better logistics.

RESIDENTIAL AND SERVICES SECTORS

The objective of the National Programme for Energy Efficiency in Buildings is to reduce the national GHG emissions by 0.9 percentage points in 2010. The programme is implemented from 2001 to 2004 with a total budget of €4 million. The key activities are the revision of the Regulations on the Characteristic of the Thermal Behaviour of Buildings (RCCTE) and the Regulations on Energy Systems and HVAC in Buildings (RSECE); the preparation of energy certification for buildings, including preparing the necessary manuals and software as well as providing training; different promotional activities; international collaboration and promotion of solar water-heating in buildings (see Chapter 7).

The RCCTE came into force in 1991. These regulations aim at improving comfort and quantifying energy demand in buildings. The requirements in RCCTE are being made more stringent and other measures, such as mandatory energy certification of buildings, are taken to strengthen the impact of the regulations. In November 2003, an expert committee proposed to make the thermal requirements for the building envelope 40% stricter than existing standards in new residential buildings. New non-residential buildings will be required to exceed the average characteristics of the existing buildings. Legislation for both RCCTE and certification is expected in 2005. The certification is expected to start in 2005 for new apartment buildings and new large non-residential buildings and to be later extended to all buildings, including the existing ones.

In 1998, the RSECE set the rules for optimising the size of air-conditioning systems to facilitate high energy efficiency. The rules are being updated as a response to the EU Directive 2002/91/EC on the energy performance of buildings. Following an expert committee report in November 2003, a new law is expected in 2004. The proposed new regulations represent a 25% improvement in minimum energy performance requirements. The new RSECE introduces strict indoor air quality requirements as well as mandatory maintenance requirements and periodic inspections for systems exceeding 12 kW. Apart from the new RSECE, no other action is taken to regulate the use of air-conditioning.

The inspections for compliance with the new regulations and for the building certification will be conducted by companies following certain quality systems or certified individuals. The verification process for compliance with the regulations will be ensured both in the application process for the building permit and for the use permit once the building is ready, whereas up to now, the verification has taken place only in the application of the use permit.

All steam boilers with a total heating surface of over 100 m^2 must have periodic test of their energy efficiency and emissions.

Energy efficiency labelling is required for refrigerators, washing machines, dryers and washer-dryers, and air-conditioners and electric heaters for household use. The EU directives on energy efficiency requirements for refrigerators, freezers and fluorescent lighting have been adopted.

Also, service sector establishments have been voluntarily conducting energy auditing while not being subject to specific regulations, as is the case for industrial enterprises. Audits have been conducted, for example, in hotels.

CRITIQUE

The government projects energy consumption in industry, transport and residential/commercial sectors to grow by 11%, 20% and 14% respectively from 2002 to 2010 while the growth in these sectors between 1990 and 2002 has been 24%, 80% and 61% respectively. These figures seem to be overly optimistic, in particular for the transport sector where passenger transport volumes are expected to increase by 42% and freight transport volumes by 38% between 2000 and 2010.

It is a matter of concern that throughout the 1990s energy demand has grown faster than GDP, therefore increasing energy intensity, while on average energy efficiency has been constantly improving in IEA member countries. At the same time, there have been few measures in place to address energy efficiency as compared to most other IEA member countries. While this is partly attributable to the improving living standards, there is ample potential in Portugal to improve energy efficiency. Though significant new action has been taken recently to strengthen the policies, a more solid energy efficiency strategy is imperative. This should include both national and sectoral targets with thorough monitoring. The policies and measures to be included should be chosen on the basis of their cost-effectiveness. In doing so, Portugal should seek every cost-effective possibility for energy efficiency improvements as well as fully implement the measures in the EU framework. To date, cost-effectiveness of the measures has been evaluated only to a limited degree and it is not clear how the results have been transposed to policy-making.

The Management Regulations for Energy Consumption is the principal policy to save energy in industry, a sector with probably a great potential for energy conservation. The government assesses that this policy has contributed to a 3% reduction of energy consumption of audited plants, which is a surprisingly small outcome compared to the achievements in many other voluntary programmes in other countries. However, it is not necessarily clear whether this has been an "additional" reduction driven by this policy or one which should have been achieved in any case through normal investment in replacing old equipment. Furthermore, small and medium-sized enterprises accounting for the majority of industrial sectors are not covered under this scheme.

The introduction of the EU-ETS could give a good incentive to large emitters to further improve their energy efficiency. However, to address the large number of small and medium-sized enterprises not covered by the EU-ETS and of large emitters not participating in the EU-ETS, a new type of arrangement would be necessary. Combining the voluntary agreement supported by incentives, such as reimbursement of CO_2 tax, could be an option. Preparation of such mechanisms with clear benchmarking should be accelerated.

In the residential and services sectors, the labelling of appliances is a good way to inform the consumer about efficient energy use. Efficiency standards, building certification and regular inspections of boilers and air-conditioners can all increase energy efficiency in these sectors. This could help in curbing the expected growth of electricity demand for air-conditioning. However, the requirement for new non-residential buildings to exceed the average thermal efficiency of the existing building stock appears to be quite weak and could be further strengthened.

Energy demand in the transport sector has been increasing faster than economic growth. Decoupling demand growth from economic growth is one of the key measures used in transport policy to curb demand. There are various possibilities to implement the de-linking and they should be fully explored. The options include reducing passenger and freight transport activity growth, modal shift towards more energy-efficient modes such as rail transport or public road transport, and improving the fuel efficiency of the car fleet.

Transport policy should contribute to energy and environmental policy objectives through, for example, better traffic management and decreasing empty trips by trucks while improving transport services through reduction of congestion. The policy to avoid empty trips by trucks is a particularly good one which should be evaluated. If it is successful, the results would also be interesting for other countries. It appears that there is considerable room for improvement of interdepartmental co-ordination among the different ministries responsible for transport, private cars, taxation and energy policy.

The Portuguese transport system relies heavily on private road transport while the share of rail and public road transport has been decreasing in recent years. The government wishes to reverse previous trends and is making efforts in terms of infrastructure investments in urban systems. However, it is not yet clear that these measures will suffice to bring the trend of rising private road transport to a halt. It is, therefore, important to foster the more energy-efficient modes. Such efforts include improving railway services and encouraging the use of public transport. Better co-ordination with municipalities or other regional organisations could enhance the possibilities of energy policies to improve energy efficiency in transport.

Regarding private road transport, car ownership is expected to continue growing significantly in the coming years. It is, therefore, important that new cars are as efficient as possible. The European agreement with the car industry to bring on the market more fuel-efficient cars should be valorised in such a way that the car fleet will increasingly consist of such fuel-efficient cars. This should be supported through active government policy, including labelling, purchase and circulation tax incentives and schemes that take the older and less fuel-efficient cars out of the market (for example, scrapping premiums). It is encouraging that the government has started considering and implementing some policies in this direction and such efforts should be further strengthened as this would also improve local air quality.

Given the level of challenge in the transport sector, even stronger measures should not be precluded. These include fuel and CO_2 taxes, and road pricing.

RECOMMENDATIONS

The government of Portugal should:

- *Establish an integrated national energy efficiency strategy as soon as possible. This should incorporate, at the national and sectoral levels, targets and strong cost-effective measures, including full implementation of the measures in the EU framework.*
- *Evaluate the cost-effectiveness and achieved energy savings of the energy efficiency projects.*

- *Enhance efforts to address the sectors not covered by the EU emissions trading scheme by, for example, the CO_2 tax and reimbursement in the case of taking appropriate measures such as more effective voluntary agreements and measures in the building sector.*

- *Curb energy demand growth in the transport sector by:*

  - *Decoupling transport activity growth from economic growth, considering a wide range of policies such as better urban planning, promoting teleworking, road pricing and modernisation of the economy away from transport-intensive activities.*

  - *Fostering more energy-efficient modes such as railways.*

  - *Encouraging car buyers to choose fuel-efficient cars and to retire old and inefficient cars by economic and regulatory measures, in particular vehicle taxation.*

GALP ENERGIA

In 1999, the State created a new holding company, GALP Petróleos, which combined all the operations of Petrogal (oil company), Gás de Portugal (GDP, gas distribution) and its subsidiary Transgás (gas import, transmission and sales company). In 2000, the new company was named GALP Energia. The owners of GALP Energia were the State 34.8%, ENI 33.3% (Italian oil company and utility), Electricidade de Portugal 14.3%, Caixa Geral de Depósitos SA (Portuguese public bank) 13.5%, Iberdrola SA (Spanish utility) 4%, and others 0.1%. On 31 March 2004, the government signed final agreements with ENI and Iberdrola on the restructuring of GALP Energia. GDP will be eventually owned by EDP (51%) and ENI (49%), although ownership will be first transferred to EDP, ENI and the electricity transmission operator, Rede Electrica Nacional (REN) for a maximum period of 18 months. ENI exchanges its shares in GALP Energia to those in GDP. Each company will pay GALP Energia about €400 million for the natural gas assets that GALP Energia owns and they will now become part of GDP. GALP Energia will be left with oil assets only. GDP's gas transmission operations will be merged with REN. Iberdrola will sell its stake in GALP Energia and pick up shareholding in two small gas distribution companies.

OIL

INDUSTRY STRUCTURE

Despite full oil market liberalisation, the Portuguese oil sector is still dominated by GALP Energia's Petrogal SA. It owns the two refineries in the country and accounts for about half of the distribution market.

The two GALP refineries are in Sines and Porto. The Sines refinery produces only fuels and has a capacity of 10.8 Mt per year and the Porto refinery produces fuels, lubricants and aromatics with a total capacity of 4.4 Mt. In 2002, the utilisation rate of the refineries was 86%. The adaptation of the EU Auto Oil Programme obliged the refining industry to make investments to meet with the new specifications for gasoline and diesel for 2000 and 2005. However, derogations to the sulphur limit for diesel and gasoline was granted to Portugal up to January 2001 and January 2002, respectively. In addition to tightening environmental regulations, the need to change the mix of productions, notably from gasoline to diesel, necessitates new investments. No disruption or problems were observed during the upgrading of the refineries. The investments in new equipment during the past few years have been €180 million at the Sines refinery and €95 million in Porto.

In 2002, eight distributors, namely GALP, Shell, BP, Repsol, Cepsa, Exxon, Total and Agip, accounted for 96% of total oil product sales. The share of GALP in gasoline and diesel sales was 40%, LPG 45% and fuel oil 59% in 2001. The number of filling stations has decreased in the last year, mainly because safety legislation imposed the closure of about 250 stations operating under residential or commercial buildings. Some closures have also taken place following rationalisation made by the distribution companies. At present, the total number of stations is about 3 050, of which GALP operates 900. The number of self-service, and to some extent non-manned stations, has increased representing 10% of the total. Licences for the opening and operation of service stations on main roads are given by the regional delegations of the Ministry of Economy and in the urban areas by the municipalities.

SUPPLY, DEMAND AND TRADE

Oil supply increased from 11.7 Mtoe in 1990 to 16.4 Mtoe in 2002. The share of oil in TPES decreased between 1973 and 2002, from 75.4% to 62.1%. This is still higher than the IEA average of 41% (2001). The share of oil is projected to decline to 54% by 2010 mainly because of the increased use of natural gas and renewables in electricity generation.

Out of the total oil supply in 2002, transport had the largest share (42%), followed by industry (27% including non-energy use) and power generation (15%). The residential, services and agricultural sectors accounted for 11% of total oil use. Diesel dominates road transport energy consumption with a share of 62%, which is much higher than the IEA average of 37% in 2001. Gasoline consumption has increased by almost half, and that of diesel in the transport sector has more than doubled between 1990 and 2002. GALP expects gasoline demand to increase slowly by about 1% per year for the next five years and decline thereafter, whereas a continuous growth of 2% per year is expected for diesel demand for several years to come. Despite some increase in the use of natural gas and combustible renewables and wastes, oil is still the main fuel (52.6%) used in the industry sector.

Portugal holds no discovered commercially viable oil reserves and all oil is imported. Some 40% of oil imports are based on long-term contracts.

Figure 11

Final Consumption of Oil by Sector, 1973 to 2010

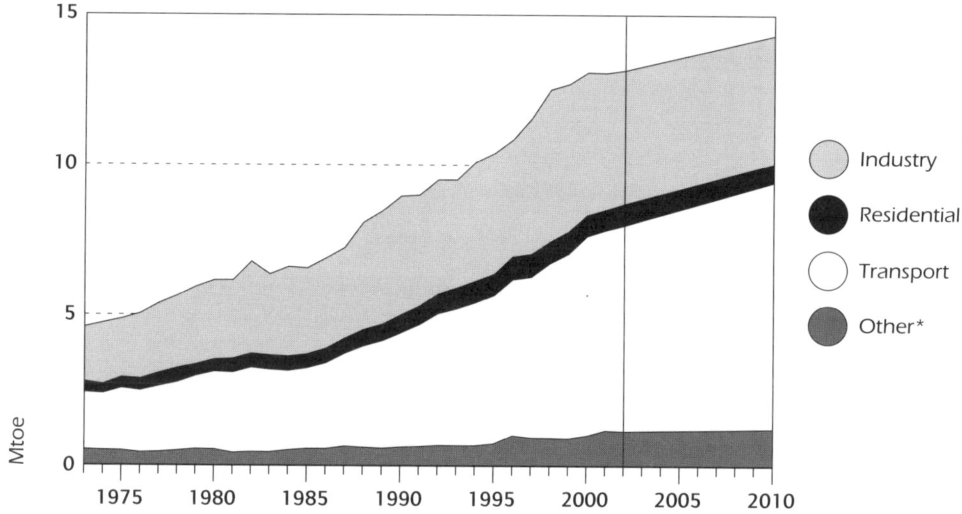

* includes commercial, public service and agricultural sectors.

Sources: *Energy Balances of OECD Countries*, IEA/OECD Paris, 2003; and country submission.

Crude oil imports totalled 11.5 Mt and product imports 5.2 Mt. Product exports were 1.4 Mt. Crude oil imports came from diverse sources, the largest being Nigeria (23%), Saudi Arabia (16%), the UK (10%) and Brazil (9%). 75% of oil products were imported from diverse OECD countries. Also the majority of exported oil products were sold to the OECD markets.

PRICES

Gasoline and diesel prices are in the mid-range within IEA member countries (see Figures 12 and 13). However, the pre-tax prices are somewhat at the higher end.

The government had applied a system of administrative prices for oil products with price ceilings, but these have been progressively lifted and finally abolished. The last remaining price ceilings were applied for gasoline and automotive diesel, including for agricultural use, until the end of 2003. When these ceilings were removed, no significant movement in market prices was observed.

Figure 12

OECD Unleaded Gasoline Prices and Taxes, Third Quarter 2003

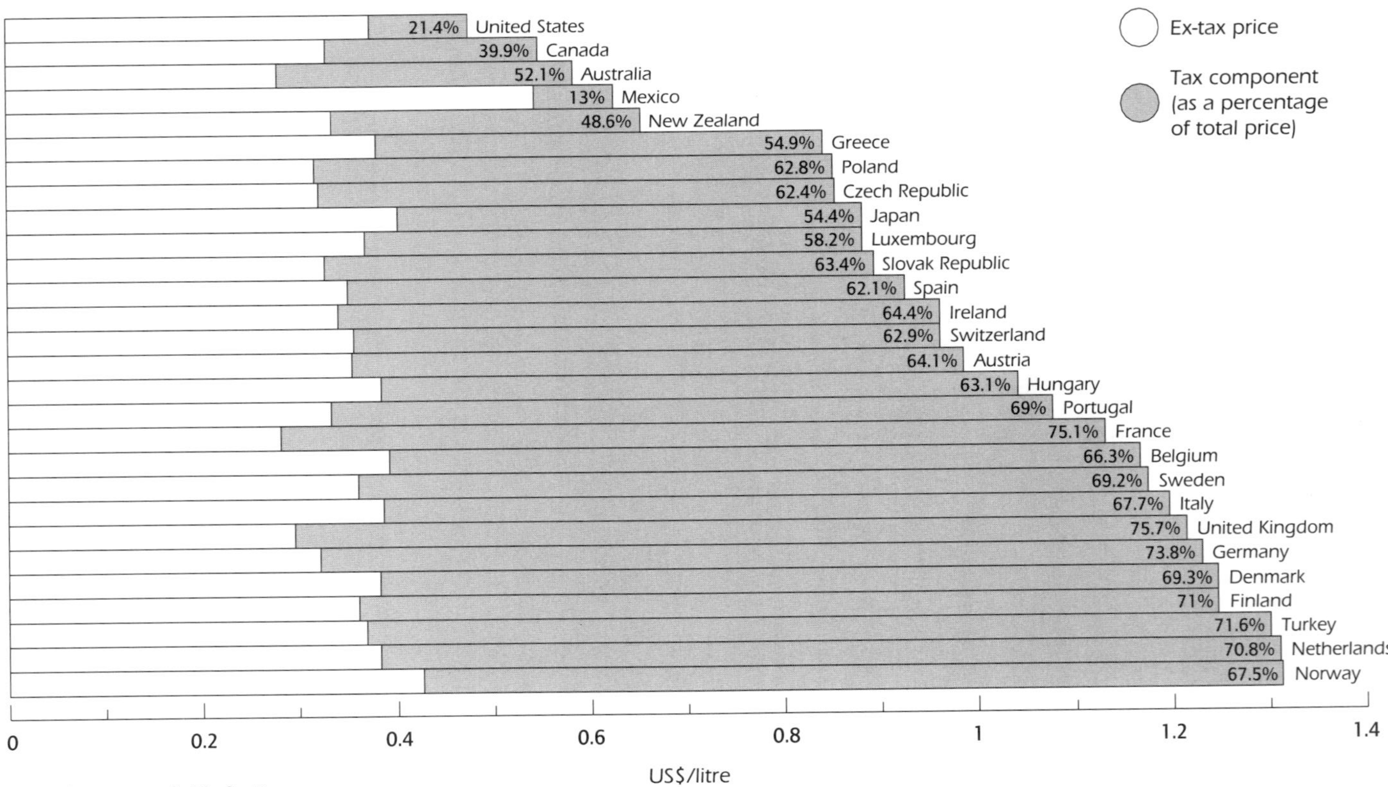

Note: data not available for Korea.

Source: *Energy Prices and Taxes*, IEA/OECD Paris, 2003.

Figure 13

OECD Automotive Diesel Prices and Taxes, Third Quarter 2003

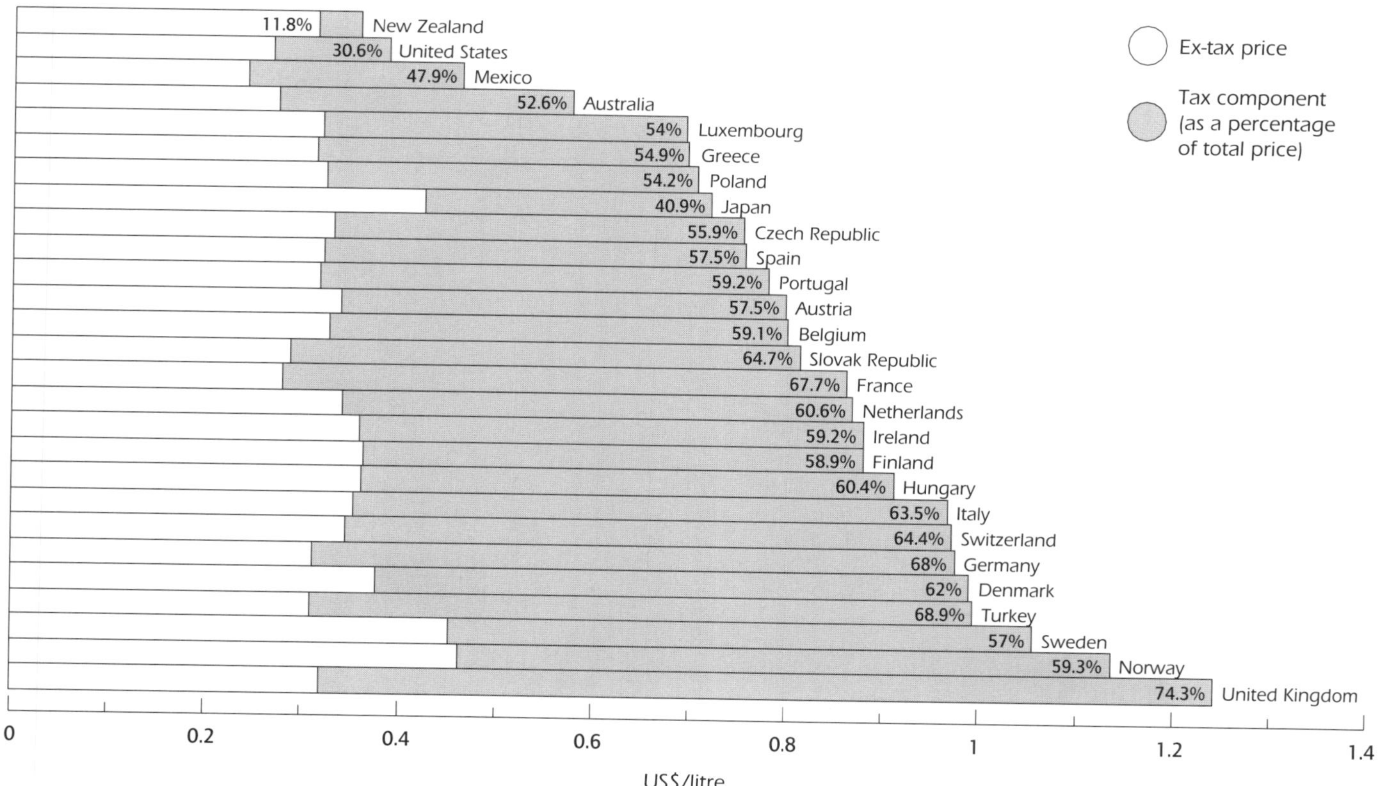

Note: data not available for Canada and Korea.

Source: *Energy Prices and Taxes*, IEA/OECD Paris, 2003.

Figure 14

Fuel Prices, 2002

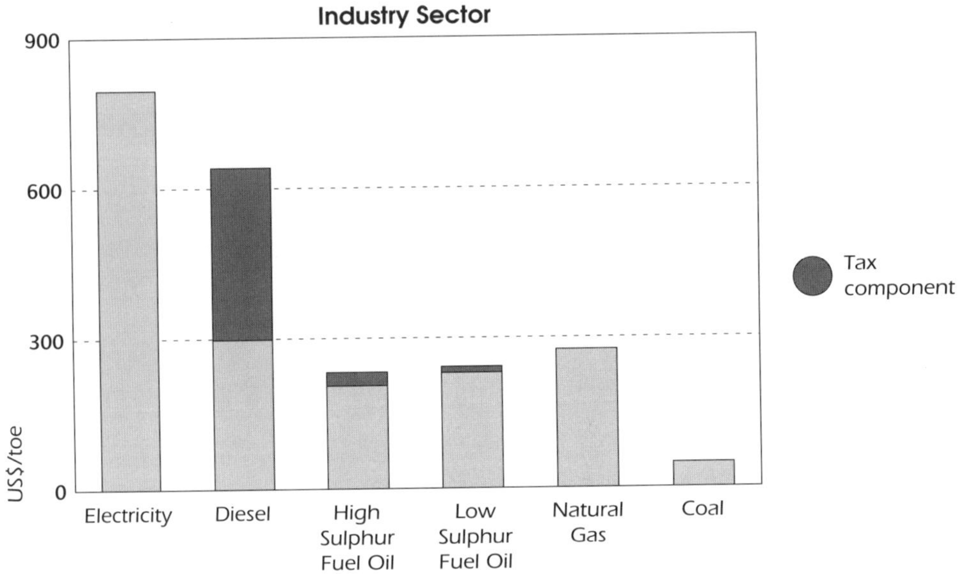

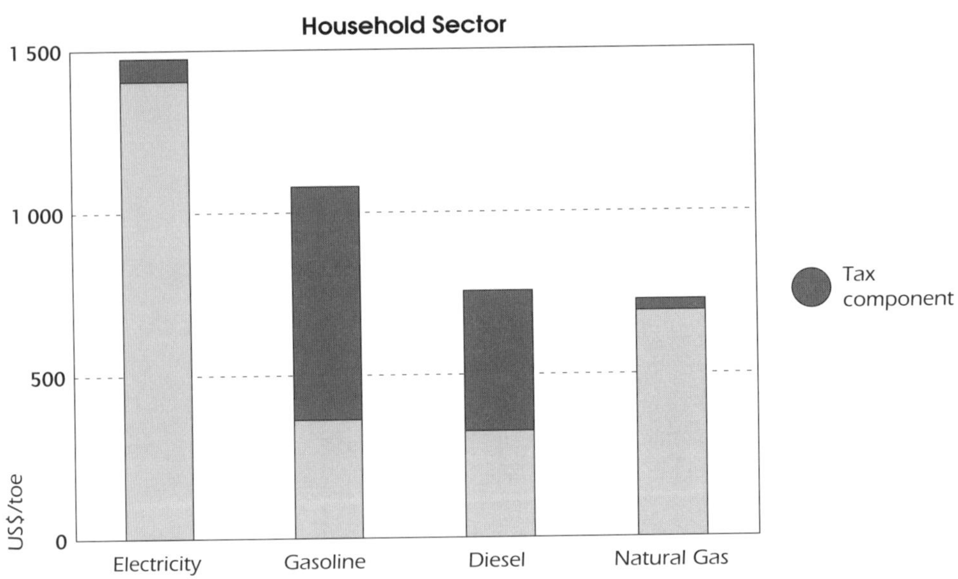

Source: *Energy Prices and Taxes*, IEA/OECD Paris, 2003.

EMERGENCY RESPONSE MEASURES

Because of the country's high dependence on oil, the Portuguese Administration has attributed great importance to oil supply security very early and in 1937 imposed an obligation to hold emergency oil stocks (which remained valid in its core until 2001). This obligation stipulated that one-third (approximately 122 days) of the previous year's consumption of gasoline, gas oil and fuel oil had to be held by the oil industry as emergency reserves. When joining the IEA in 1981, Portugal already held consecutive amounts of emergency oil stocks.

However, in recent years, Portugal has frequently not been able to comply with its IEA obligation to hold oil emergency reserves of 90 days of net imports. Inconsistencies in the national legislation with the International Energy Program (IEP) stockholding obligation and an increasing lack of stockholding capacity contributed to Portugal's non-compliance. Consequently, the Administration undertook major efforts to refurbish the Portuguese emergency legislation and to restructure its stock obligation. The oil companies constructed some new storage capacity and renewed some of their existing storage facilities.

The National Emergency Sharing Organisation (NESO) is the central body for oil supply emergency co-ordination. As suggested by the 1996 Emergency Response Review Team, its structure was significantly simplified in 2002. Generally, the organisation became centred on the General Directorate for Geology and Energy (DGGE) within the Ministry of Economy, overtaking the responsibilities of the former structure called *Organização para Emergência Energética* (OEE). At the same time, emergency response competences and measures became better defined and legally supported. In addition to the DGGE, the NESO is composed of the *Comissão de Planeamento Energético de Emergência* (CPEE) which, in the case of a crisis, evolves into the *Conselho Nacional de Emergência Energética* (CNEE). The activation of the NESO is foreseen in Decree-Law 114/2002 in an emergency situation; it may also be activated through the legislation of the Civil Emergencies Planning Board under the authority of the Minister of Defence.

In 2001, new legislation concerning specifically the emergency stockholding obligation was passed (Decree-Laws 10/2001 and 339D/2001), conforming to EU requirements and adopting some IEA rules (namely a 10% deduction on reserves counting). The oil industry has the obligation to hold emergency oil stocks of the three EU product categories equivalent to 90 days of the previous year's consumption. It is also obliged to hold 30 days stocks of LPG. The laws further set the rules for a planned stockholding agency, the *Entidade Gestora de Reservas Estratégicas de Produtos Petrolíferos* (EGREP), which was to hold at least one-third of total

Portuguese emergency reserves on behalf of, and financed by, the oil industry. After some start-up difficulties, this agency was finally established in December 2003, with an ambitious stockholding schedule for 2004.

Further, as storage capacities recently represented a constraint on oil storage for the industry, a governmental decree (Decree-Law 71/2004) to allow EGREP to hold stocks in other EU member States was published in March 2004. The Portuguese Administration has started consultation with several EU member States in order to hold stocks abroad as soon as possible. However, the emergency oil stocks were still only 85 days of net imports on 1 April 2004, will below the 90-day obligation.

COAL

In 2002, coal demand was 3.5 Mtoe, 26% more than in 1990. Coal's contribution to TPES was 13.2%, lower than its 15.5% share in 1990. Almost all coal is steam coal used for electricity generation and in the cement industry. Imports of coking coal ceased along with the shutdown of the blast furnaces in the iron and steel industry. All coal is imported. The main sources are South Africa (40%), Colombia (25%), Australia (15%) and Indonesia (11%). The government estimates coal demand to decline to 3.1 Mtoe by 2010 but considers it important to use some coal in power generation to enhance security of supply via a balanced generation mix.

Figure 15

Final Consumption of Coal by Sector, 1973 to 2010

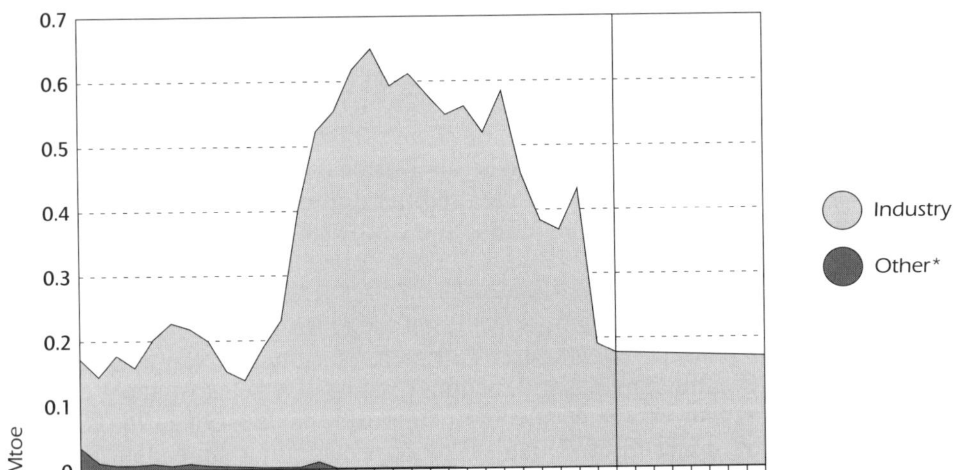

* includes commercial, residential, public service and agricultural sectors.

Sources: *Energy Balances of OECD Countries*, IEA/OECD Paris, 2003; and country submission.

NATURAL GAS

INDUSTRY STRUCTURE

The dominant player in the gas market is GALP Energia through its subsidiaries, Transgás and Gás de Portugal Distribuição (GDPd). Details about GALP's shareholding structure and restructuring are given at the beginning of the chapter. Transgás is the supply and transmission company but it also sells gas to consumers with an annual consumption exceeding 2 mcm. It has currently an exclusive concession contract for 35 years to import and store gas.

GDPd runs GALP's distribution businesses. It has a stake in all the six regional distribution companies, namely Lisboagás (100% in 2001), Portgás (46.6%), Lusitâniagás (84.1%), Setgás (45%), Tagusgás (40.9%), Beiragás (58.6%). In addition, GDPd has set up four autonomous gas distribution units called *Unidades Autónomas de Distribuição de Gás* (UADs) to serve small towns. In 2003, 18% of total gas supply was sold through the distribution companies with the majority sold directly to large consumers. All regional distribution companies receive their gas from Transgás on long-term contracts of 25 years.

INFRASTRUCTURE AND SUPPLY SOURCES

Natural gas imports to Portugal began in February 1997. In 2002, 87% of gas was imported via pipelines from Algeria under long-term contracts. The remaining 13% was LNG imported mainly from Nigeria. The imports from Nigeria have been delivered through the Spanish terminal at Huelva and piped to Portugal.

Possibilities to import LNG have been substantially improved recently as Portugal's first LNG terminal was commissioned at the end of 2003 in Sines. Commercial activities started in January 2004 with imports from Nigeria, and were based on new long-term take-or-pay contracts. The initial capacity of the terminal was 2.6 bcm per year but has already been extended to 5.2 bcm. The capacity to emit is 600 000 normal cubic metres per hour (Ncm/h), with the possibility to reach 900 000 Ncm/h in peak hours. Its two tanks have a total storage capacity of 210 000 cm of LNG. The maximum size for LNG vessels to be received and unloaded is 165 000 cm.

In the mid to longer term, Transgás aims at importing half of its supplies via pipeline and half as LNG. The objective of this division is the reduction of gas supply costs with improved security of supply as a side-benefit. At present, the pipeline gas contract is for 2.3 bcm (started in 1997 for 25 years) and the three LNG contracts from Nigeria are for 0.42 bcm (started in 2000 for 20 years), 1 bcm (started in 2003 for 20 years) and 2 bcm (starts in 2005 for 20 years).

The maximum import capacity from Algeria by pipeline is 4.5 bcm per year. There is a possibility to increase the capacity of the Europe-Maghreb line by 4 bcm with relatively low investments in compressor stations and by a further 4 bcm by investing heavily on the pipelines.

The total length of high-pressure networks is over 2 000 km and medium and low-pressure networks almost 9 000 km. The growth of networks has clearly exceeded GDP's forecasts presented in 1999 for 2010. The Portuguese network is connected to the Spanish network in two locations, in the north and in the east (see Figure 16).

The construction of underground storage with an initial capacity of 165 mcm is under way in salt caverns in Carriço. The first two cavities with a capacity of 90 mcm were commissioned in early 2004, the third will be commissioned in 2006 and the fourth by 2007. The site allows for the construction of up to 10 cavities.

On average, 36% of the investments for different gas infrastructures have been financed by the EU. Between 1994 and 1999 the investments totalled €485 million. At present, the subsidies are given in the framework of the PRIME Programme (see Chapter 3). Table 5 shows PRIME's subsidies for energy infrastructures, principally natural gas. The EU also provides additional grants for promoting the increased use of natural gas, in practice for GDP's marketing activities. In 2003, this support totalled €145 000.

Table 5

PRIME Support to Energy Infrastructures[1]

(million euros)

Year	Total budget	Public expenses				
		Total	Support from EU	National public financing		
			Total	Total	Central admin.	Other public
2000	49.53	49.53	14.63	34.90	3.66	31.24
2001	50.41	50.41	13.91	36.49	3.49	33.01
2002	33.72	33.72	13.01	20.71	3.26	17.45
2003	23.85	23.85	12.52	11.33	3.13	8.20
2004	18.47	18.47	10.36	8.10	2.59	5.51
2005	19.24	19.24	10.67	8.56	2.67	5.89
2006	20.44	20.44	11.15	9.28	2.79	6.49
Total	**215.65**	**215.65**	**86.26**	**129.39**	**21.60**	**107.79**

[1] The budget is indicative. It is mainly composed of support to gas infrastructures but small amounts for electricity networks for connecting renewables are included.

Source: The Ministry of Economy.

Figure 16

Natural Gas Infrastructure

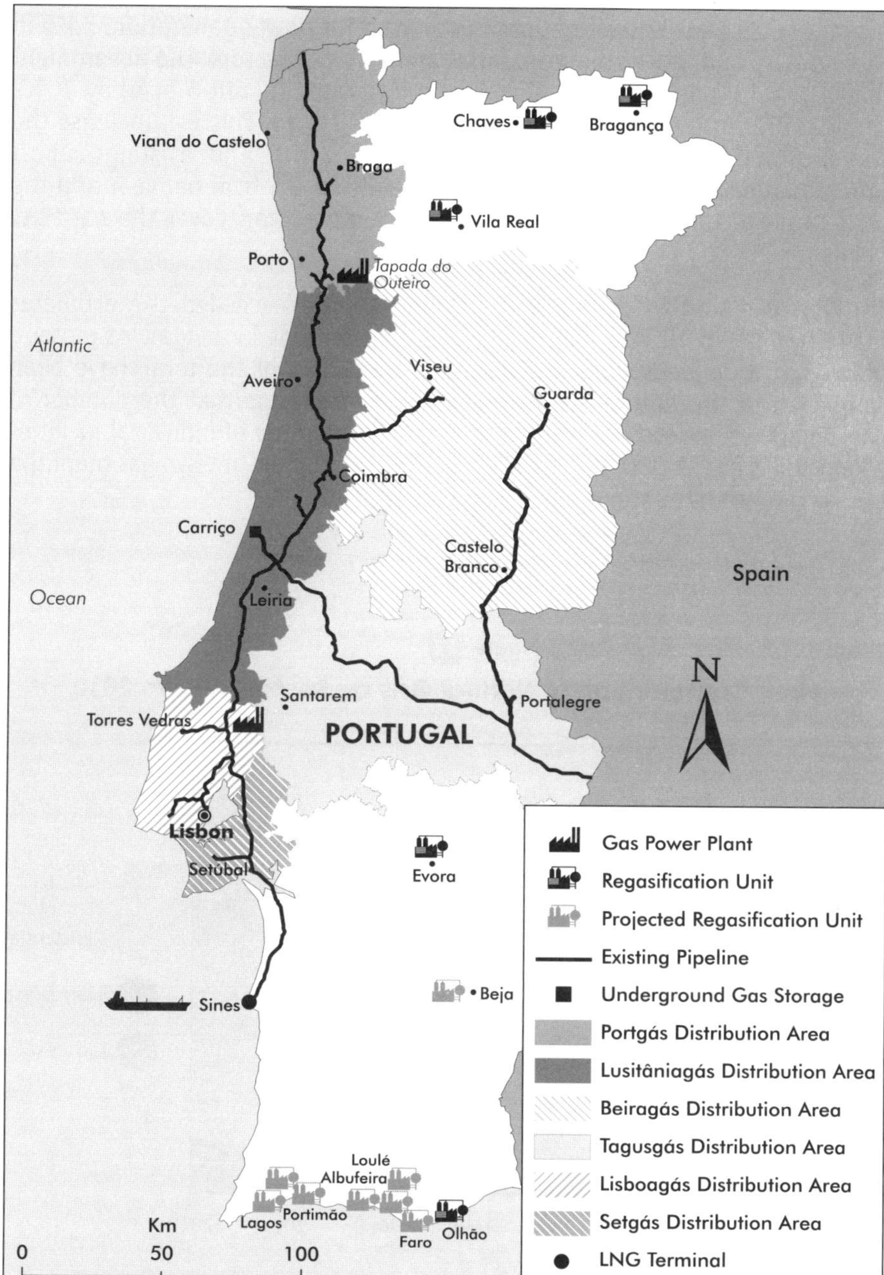

Source: *Natural Gas Information 2003*, IEA/OECD Paris, 2003.

DEMAND

The Portuguese natural gas market is new. Gas demand was 2.7 Mtoe (3.1 bcm) in 2002. It represented 10.3% of TPES whereas the IEA Europe average is 23.4%. 57% of natural gas is used for power generation, 34% in the industry and 9% in the residential and services sectors. The government forecasts total gas demand to increase to 5.1 Mtoe (about 6 bcm) by 2010, whereas Transgás's forecast is a bit higher, at 7 bcm. This will increase the share of natural gas in TPES to 17.1% in 2010. The existing import infrastructures, including the pipeline capacity of 4.5 bcm per year and the LNG capacity of 5.2 bcm per year, already more than cover the capacity needs for 2010.

The volumes sold to distribution consumers have exceeded the estimates made by GDP in 1994 in all distribution areas except for Setgás. At present, about 750 000 households, representing over 15% of the total, have been connected to the natural gas networks. GDP estimates that the number of consumers will exceed one million in 2007. The number of industrial facilities connected is over 2 000. In Lisbon, converting consumers of town gas (naphtha cracker) to use natural gas was completed in 2001.

Figure 17

Final Consumption of Natural Gas by Sector, 1973 to 2010

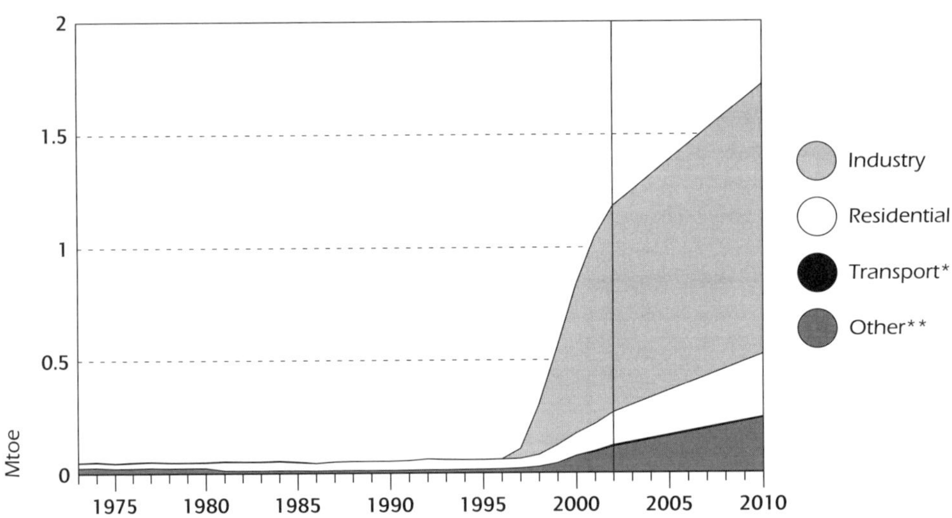

* negligible.

** includes commercial, public service and agricultural sectors.

Sources: *Energy Balances of OECD Countries*, IEA/OECD Paris, 2003; and country submission.

SECURITY OF GAS SUPPLY

Decree-Law 13/2001 regulates emergency situations for energy, including natural gas. The policies applied to security of supply problems include the use of long-term contracts, diversification of supply sources, storage obligations, use of interruptible contracts, co-operation with Spain and monitoring of investments.

Both short- and long-term security of gas supply is enhanced through the concession contracts. The concession contract for the transporter, namely Transgás, imposes a stockholding obligation of 20 days of average daily non-interruptible consumption based on the consumption in the previous year. Gas used for power generation in the Tapada do Outeiro power plant is excluded from this obligation owing to its multi-firing possibilities. Because there is no gas production, Transgás has built an underground storage to supplement line-pack and the storage capacity in the LNG terminal. The LNG terminal was built by Transgás, obliged by the concession contract, to avoid total dependence on a single supplier (see Infrastructure and Supply Sources above).

Transgás and the Spanish transmission company Enagas have an agreement on a shared strategy to develop and integrate the peninsular natural gas grids. This is implemented by creating joint companies which will run the pipelines supplying gas from Algeria to Spain and Portugal and to the northern region of Spain through Portugal. The agreement was introduced to simplify the financing of cross-border gas deals.

The administration monitors the investments through periodic monitoring reports from the gas companies. The concession contracts contain enforcement instruments related to investments such as minimum investment obligations for networks expansion.

PRICES

To promote the use of natural gas, there are no excise taxes and VAT was set at 5%.

Prices for large industries, which are indexed to the price of heavy fuel oil, are revised on a monthly basis. Transgás sells natural gas to large industries at a price comparable to that of heavy fuel oil, which makes it very competitive with propane. There are volumes-related rebates and in 2002, the discounts were about 10%. Interruptible consumers represented 4% of total gas sales in 2002.

Transgás sells gas to Turbogás's combined-cycle gas turbine (CCGT) plant under a long-term take-or-pay contract. Calculations for gas tariffs are negotiated every two years.

Tariffs for small industrial clients, with a consumption of 0.01-2 mcm per year, have no reference formula in the concession contracts. However, concession contracts forbid price discrimination. Prices are revised every three months. In this sector, prices for natural gas are lower than those for propane, slightly higher than for heavy fuel oil and clearly higher than for solid fuels.

The base for tariff calculation is set in the concession agreements and differs among distributors. Tariff calculations for residential and commercial consumers have a fixed and a variable part. The variable part (commodity-related) is revised every three months according to the variation in Transgás's prices. On average, the price of natural gas for the residential sector is lower than the price of electricity, piped propane and LPG.

According to Eurostat (see Figure 18) in July 2003, prices in Portugal for the very smallest household consumers (200 cm per year) were 5.3% lower than the EU average. They were 3.8% above the average with an annual consumption of 400 cm and 6.7% above the average with an annual consumption of 2 000 cm. For large consumers, prices were 11.3% above the EU average for consumers of 0.1 mcm per year, 2.9% below the average for consumers of 1 mcm per year with a 1 600 hour load factor and 1.6% above the average for consumers of 1 mcm per year with a 4 000 hour load factor.

MARKET REFORM

The 1998 EU Gas Directive allowed a derogation for Portugal from market liberalisation until 2008-2012 (depending on the consumer group) because it was considered an emergent gas market. However, the government intends to liberalise the natural gas sector earlier because it recognises the benefits of increased competition as well as the importance to co-ordinate the convergence of the electricity and gas sectors in the Iberian market.

The first phase of market opening has been announced. The Resolution of the Council of Ministers 63/2003 indicated that electricity generators, currently representing half of total natural gas sales, will be able to choose their supplier from 1 July 2004 in line with the new EU Gas Directive. However, as the new directive has not yet been transposed into national legislation, it has not been defined how this will be implemented.

The body responsible for industry regulation is the Energy Services Regulatory Authority (ERSE) which had its role extended to natural gas activities in early 2002.

Some details of the new gas market design are emerging. In line with the new 2003 EU Gas Directive (2003/55/EC), supply and transmission activities will have to be separated. This will be implemented by the creation of an independent transmission system operator (TSO) via ownership unbundling, going one step further from the legal unbundling required by the directive. The Resolution of the Council of Ministers 63/2003 indicates that Trangás

Figure 18

Gas Prices in Portugal and in Other Selected Countries, 1 July 2003

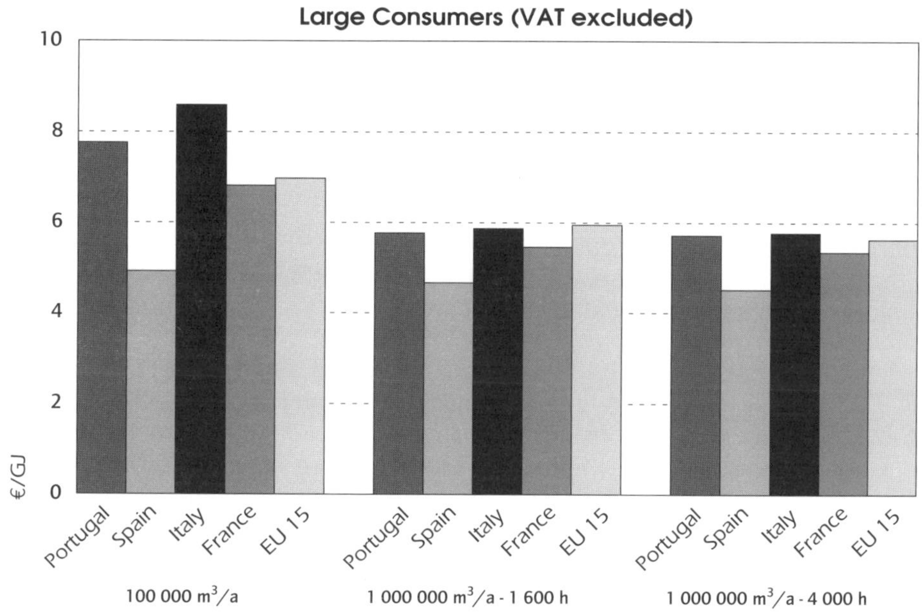

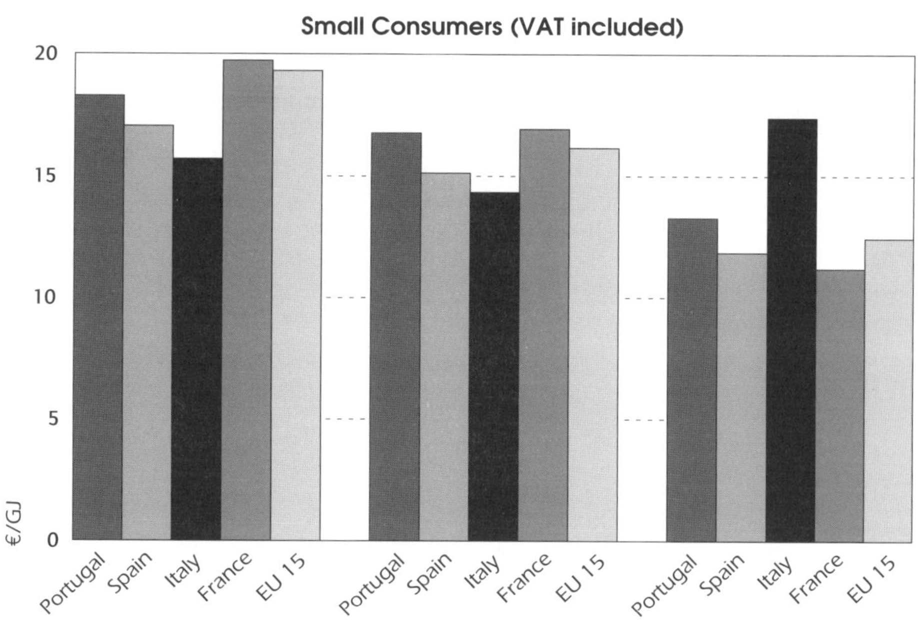

Source: Eurostat.

will be separated from GALP Energia and integrated into REN, the independent electricity TSO. This is similar to the British organisation for NGT Transco, which includes both the electricity and gas grids. It is envisaged to privatise the new TSO. Distribution companies will be subject to legal unbundling of retailing and distribution.

Third-party access (TPA) to the gas networks and the LNG terminal will be based on regulated tariffs, in line with the 2003 Gas Directive. TPA will also be introduced for underground gas storage. All TPA tariffs will be established by ERSE.

At present, the gas sector is regulated through concession contracts. It has not yet been established how this will be dealt with in the process of earlier liberalisation to minimise stranded costs.

One important aspect in market liberalisation will be consumer protection. In June 2002, the General Directorate for Geology and Energy (DGGE) issued the first regulation on the quality of services which the natural gas distributors must provide to their customers. The regulation establishes the minimum technical and commercial standards for the gas transport and distribution operators that are subject to public service obligations: the operators must install measurement, recording and monitoring systems that allow for the evaluation and verification of the accomplishment of quality of service standards defined in the regulation. The technical standards include those related to the continuity of service, such as the number and duration of the scheduled and unscheduled interruptions, and to the properties of the gas supplied. The commercial standards regulate the relations of the operators with their clients, such as compliance with clients' requests, information, technical assistance and evaluation of clients' satisfaction.

The Portuguese and Spanish governments are working on extending the integrated Iberian electricity market, MIBEL, to include gas. The project is, however, at a very early stage and the details of its implementation are not yet known.

CRITIQUE

OIL

Total oil demand has continued to increase, driven particularly by the transport sector. Annual variations, however, depend largely on hydropower generation. The increasing use of diesel in transport and expected decline in gasoline use will put pressure on the refineries to adapt their processes to accommodate these trends. Up to date, adequate investments have been made but diesel imports may increase in the future.

With the removal of the remaining price ceilings, the oil sector has been fully liberalised. This is a welcome development. It is also notable that the removal of price ceilings took place without any significant impact on prices, which is an indication of a functional market. However, even though GALP's share of the retail market has decreased, it still has a dominant part of the domestic market. There is no evidence that the company is using this position to hamper competition. Nevertheless, such a dominant position warrants attention by the competition authority.

With the high share of oil in total primary energy supply, oil supply is the key for energy security in Portugal. However, Portugal has not been compliant with its IEP emergency reserves obligation of holding oil stocks equivalent to 90 days of net imports since late 1992. This situation has been continuously addressed by the IEA and its members as energy security in today's energy environment depends on solidarity. Positive decisions have been taken by the government, but they still need to be fully implemented.

NATURAL GAS

Natural gas was introduced in Portugal for the first time in early 1997, the benefits of which are multiple. Natural gas will reduce Portugal's high dependence on imported oil and will continue to diversify energy supply. Increased natural gas consumption will also alleviate increases in CO_2 emissions. In this context, the fact that natural gas has been rapidly and successfully penetrating into the energy mix despite its relatively recent introduction is a positive development and the government of Portugal should be commended for its efforts.

While Portugal had been dependent on a single supply source, Algeria, the new LNG terminal has significantly enhanced security of supply both by allowing diversification of supply sources and by providing enough supply capacity for several years. Since the existing contracts of about 5.7 bcm will not be sufficient to meet the estimated demand of 6 to 7 bcm in 2010, there may be further potential of diversification, particularly if demand follows the larger estimate made by Transgás. In the mid to longer term, Transgás envisages importing half of its supplies via pipeline and half as LNG, which will contribute positively to energy security. It is also noteworthy that security of gas supply has been enhanced through the concession contracts and the construction of gas storage.

Investment subsidies for natural gas infrastructure from the government and the EU, the exemption of excise tax and the lower VAT, have contributed greatly to the rapid penetration of natural gas. The main transmission lines have been built and the pace at which Gas de Portugal (GDP) acquires new distribution clients is slowing down owing to limited demand and competition from LPG.

New demand is expected to arise mainly from power generation for which the infrastructure appears to be already in place. Therefore, the review team considers that the Portuguese gas market, particularly on the high-pressure side, is maturing to a stage where there is no need for government intervention.

It should be borne in mind that investment subsidies can result in over-investment and hence inefficient use of capital. To some extent this may have happened because the current capacity in place exceeds the projected demand, at least until 2010. Subsidies and tax exemptions will distort competition between natural gas and other fuels. Furthermore, the introduction of a domestic emissions trading scheme could provide strong incentive for increased use of natural gas in power generation, which will further reduce the necessity of other government interventions. Now that the gas market is maturing on the high-pressure side, subsidies should be limited to distribution and progressively phased out with a view to the local distribution companies' need to still develop their gas infrastructures. In the liberalised markets, investment should preferably be enhanced by a good investment climate which can be provided by solid policy, legislative, regulatory and fiscal environments.

GALP Energia has dominated the gas market from upstream to retail. It is commendable that the government has announced its intention to bring forward the beginning of the gas sector liberalisation from 2008 to 1 July 2004, even though the country has a derogation from the EU directive. It is also commendable that regulatory responsibilities for the sector have already been established with an independent regulator, ERSE. Portugal's target date for the first phase of market opening is ambitious, but not unfeasible. To achieve this, it is important to quickly establish the regulatory framework to meet this timeframe. It is also important that this framework is clear, gives sufficient powers to the regulator and provides incentives for an efficient market.

The TPA tariff for networks needs to be set. It should provide for fair and non-discriminatory access and be transparent. One prerequisite for implementing these principles is effective unbundling. The planned separation of Transgás from GALP and its inclusion into REN provides an independent TSO. It is, however, important that REN incorporates the competences on gas grid operation as well as adequate human and financial resources, as network operation is quite different between gas and electricity. REN is partly owned by EDP. Taking into account the convergence between gas and electricity and the takeover of GDP by EDP and ENI, it is important to ensure that the ownership structure of REN will not have a negative influence on the development and fair operation of the gas grids. When liberalisation advances to smaller consumers, it will also be necessary to implement at least legal unbundling of retailing from distribution, as projected.

Portugal will implement regulated TPA to the gas networks in line with the new EU Gas Directive. Implementing this, together with the unbundling of the transmission system operator, will enhance transparency. In the longer term, exemptions to third-party access can be used to stimulate investment in new facilities, particularly by new entrants in accordance with Article 22 of the 2003 Gas Directive.

Beyond the first stage, further steps of liberalisation have not been announced. This may cause some regulatory uncertainty, which makes it difficult for the market players to adequately carry out long-term strategic planning and make investment decisions. Such uncertainty could be avoided by setting a fixed schedule for the different steps of liberalisation.

The takeover of GDP by EDP and ENI will have implications for the gas sector. As the Iberian gas market will be created later than the electricity market, this raises issues regarding market power in the gas sector. EDP is not likely to have incentives to promote competition in the Portuguese gas sector.

Given the convergence between gas and electricity and the forthcoming implementation of the Iberian electricity market, the creation of the Iberian gas market appears sensible. However, its practical implementation is likely to take several years because of the very different states of development of the gas markets in the two countries. While the gas market in Spain is fully liberalised in legal terms, only the first step has been announced in Portugal. The regulatory regimes, such as access conditions and TPA tariff regulation, need to be harmonised. Natural gas prices in Spain are significantly lower than in Portugal. The Portuguese gas industry considers that it needs to clearly improve its cost structure to be able to compete with Spanish companies which may raise claims for stranded costs.

RECOMMENDATIONS

The government of Portugal should:

- *Continue to evaluate the competition situation in the oil market.*
- *On the basis of the newly established emergency legislation and the new stockholding agency (EGREP), make all necessary efforts, together with the oil industry, so that Portugal will be constantly compliant with the IEA emergency stockholding obligation in the future.*
- *Maintain the policy for diversified supply sources of natural gas and the balance between pipeline gas and LNG.*
- *Phase out subsidies and tax benefits for natural gas.*

- *Finalise the regulatory framework for the partial market liberalisation in July 2004. Set the schedule for the next steps of liberalisation.*
- *Make sure that third-party access to the facilities, including the pipelines, the LNG terminal and gas storage, is on a fair, non-discriminatory and transparent basis.*
- *Closely co-operate with Spain to introduce an Iberian gas market, preferably in pace with domestic market opening, and present a realistic time schedule for the market players.*

SUPPLY

In 2002, renewable energy supply was 3.6 Mtoe accounting for 13.8% of TPES. The share of renewable energy was higher in 1990, at 18.6%, although the absolute level was less, at 3.3 Mtoe. The supply of non-hydro renewables increased from 2.5 Mtoe in 1990 to 3 Mtoe in 2002.

There are large annual variations in the hydrological conditions. Since 1995, they have led to fluctuations in hydropower generation of as much as 30-40% per year from the average annual generation. In 2002, electricity generation from renewable energy sources was 10 TWh as compared to 16 TWh in 2001. The share of renewables in total generation was 22%, whereas in 2001 it was almost 35% owing to a better hydrological year.

From 1990 to 2002, 278 MW of small-scale hydropower plants were commissioned. Deliveries to the grid from independent hydropower producers average about 0.9 TWh. Further developing the hydro resources for electricity generation is sometimes complicated because many water systems have multiple uses such as water supply and irrigation. However, multi-purpose systems can also be a benefit to hydropower as the costs of the civil works are spread out over several uses.

Figure 19

Development of Renewable Generating Capacity

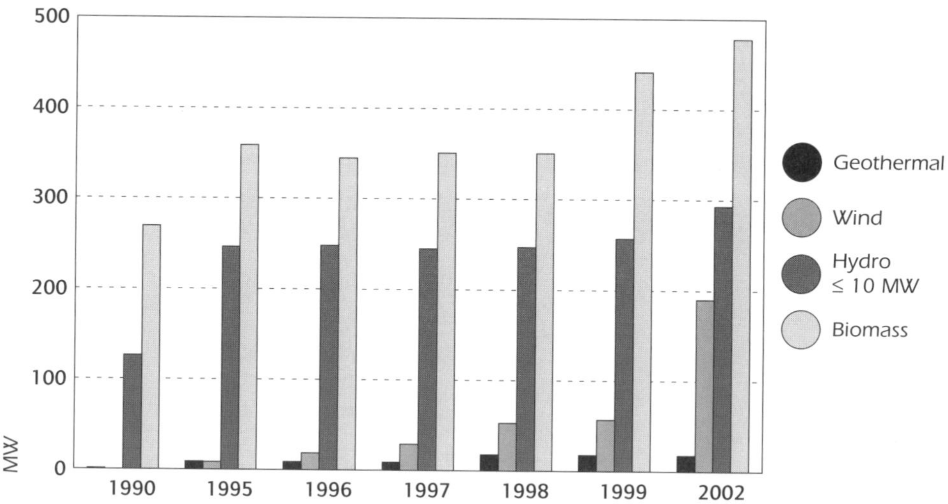

Source: Ministry of Economy.

Wind capacity was almost non-existent up to 1996 but has been increasing rapidly, reaching 101 MW in 2001 and 185 MW in 2002. Electricity generation from wind was 256 GWh and 360 GWh respectively. In some regions in Portugal the wind farms generate most of their electricity over the summer, whereas in some other regions, most of the electricity is generated over the winter.

Geothermal generation takes place on the island of S. Miguel in the Azores. Geothermal capacity increased from 3 to 9 MW in 1994 and to 18 MW in 1998. Generation from geothermal sources was 105 GWh in 2001 and 96 GWh in 2002.

POLICY

The Portuguese government promotes renewable energy to diversify energy sources and to reduce Portugal's significant dependence on imported fuels as well as for its environmental benefits. Furthermore, the government seeks larger economic benefits by developing an industrial cluster to generate jobs and to create a new export industry.

Following the EU Renewable Energy Directive 2001/77/EC, Portugal has agreed to an indicative target of generating 39% of its electricity from renewables by 2010. To reach the overall target, Portugal has established indicative targets for the different renewables in the Resolution of the Council of Minister number 63/2003 (see Table 6). The plan involves increasing the renewable electricity capacity by 5 000 MW by 2010. The necessary investments to meet the targets are estimated at about €5 billion.

As the government estimates total electricity demand to reach about 64 TWh in 2010 and large hydropower to contribute some 15 TWh, reaching the target would require almost 10 TWh to come from new wind, mini-hydro and biomass projects. In 2010, 58% of renewable electricity is expected to be generated by hydropower, 32% by wind, 9% by biomass and 0.4% by geothermal.

Renewables are promoted principally via a guaranteed market supported by feed-in tariffs and direct subsidies as discussed below. Biofuels are discussed in Chapter 4 and R&D in Chapter 9.

The transmission system operator (REN) has an obligation to buy electricity from renewable electricity generators. The purchase price is based on fixed feed-in tariffs for each renewable (see Table 7). Once a construction permit has been provided for an installation, the feed-in tariff is guaranteed for an undefined time and the level is increased in pace with inflation. Financing for the remuneration exceeding the electricity market price is collected directly from the consumers as a part of their end-use tariffs. This income transfer totalled about €28 million in 2002 but is expected to increase to €259 million by 2010. There is no mechanism in place which would automatically

Table 6

Indicative Targets for Renewable Electricity

Energy	*Installed capacity in 2001 (MW)*	*Installed capacity in 2010 (MW)*
Wind	101	3 750
Small hydro	215	400
Biomass	10	150
Biogas	1	50
Municipal waste	66	130
Wave	0	50
Photovoltaics	1	150
Large-scale hydro	4 209	5 000
Total	**4 603**	**9 680**

Source: Ministry of Economy.

Table 7

Feed-in Tariffs for Renewables, 2003

Renewable		*Feed-in tariff (€/kWh)*
Wind	2 000 hours	0.092
	2 200 hours	0.090
	2 400 hours	0.088
	2 600 hours	0.086
	2 800 hours	0.083
Small hydro	Average conditions	0.080
Wave		0.240
Photovoltaics	> 5 kW	0.240
	< 5 kW	0.450
Average "market price" for electricity		0.068

Source: Ministry of Economy.

adjust the feed-in tariffs for new projects to accommodate cost reductions (for example, improvements in technology). Instead, the government is to review the tariffs from time to time. The tariff levels were increased three times to promote renewable electricity generation over the past nine years: in 1995 (Decree-Law 313/95), in 1999 (Decree-Law 168/99) and in 2001 (Decree-Law 339-C/2001).

The feed-in tariffs have been effective in terms of capacity increases. Of the 3 750 MW indicative target for installed wind capacity in 2010, an estimated 80% has been approved following 7 000 MW of license applications registered at the beginning of 2002 after the Decree-Law 339-C/2001 was issued. Of the licensed new wind generating capacity, 500 MW is either under construction or contracts have been signed.

The Measure to Support the Harnessing of the Energy Potential and Rationalisation of Consumption (MAPE) provides support for the use of renewable energies (see Chapter 3). Its budget for renewables was €15 million in 2002 and €35 million in 2003. The corresponding total investments made were €37.5 million and €87.5 million respectively. MAPE provides soft loans with zero interest rate up to 40% of the eligible investment, excluding large hydropower.

MAPE also supports solar thermal applications. Solar energy availability in Portugal can exceed 3 000 sunshine hours per year in some areas. The objective of the Solar Hot Water for Portugal programme (AQSpP) is to reach 1 million m^2 of installed solar collector capacity by 2010. This involves increasing the annual solar collector sales 30-fold. In addition to households, which are the major market with a 66% share, new installations are also expected for pre-heating in industrial processes (25%) and in the services sector (9%), for example in sport facilities, hospitals and hotels. One of the obstacles for the programme is overcoming the "bad reputation" solar collectors earned in Portugal in the 1980s when problems were identified in all steps of the process, including manufacturing, installing and end-use. These will be addressed by long warranty periods of six years for the equipment, introducing a certification process for the installers and by extensive information dissemination. The installation of solar collectors is also promoted by providing subsidies covering 20-40% of the investment cost, tax exemptions for households, accelerated amortisation for the companies investing in solar equipment and the reduced VAT of 12%. A specific "observatory" has been established to monitor the programme.

The licensing periods for renewable energy generation appear to be long. For the 500 MW of wind power under construction, environmental licences took from 2 to 6 years to acquire from the Ministry of Environment but industry reports even longer periods. The industry considers the long licensing period a risk because financial and other conditions of the projects may change during the application process. Furthermore, the industry criticises the criteria applied for the environmental impact assessment for not being transparent.

The industry has already applied for licences for a wind capacity which in total approaches the indicative target for wind power. The government does not intend to give licences exceeding the target level owing to the difficulties and costs of connecting large amounts of intermittent wind power to the network and because of delays in the already licensed projects. In this sense, the indicative target has become effectively a cap on wind. The transmission system operator, REN, and the government see challenges in the areas of network planning and system operation. One of the approaches taken to alleviate the difficulties will be constructing new pumped storage hydropower capacity.

In Portugal, municipalities are given 2.5% of the turnover of renewable electricity plants established on their territory. Therefore, investors in Portugal have not encountered much local resistance for renewable energy projects contrary to what is increasingly faced by their peers in many other IEA member countries.

CRITIQUE

The renewable electricity target is challenging for several reasons. First, it requires more than double the existing generating capacity, particularly with regard to mobilising significant investments in several of the new renewables. Second, electricity demand is expected to continue its rapid growth, further adding to the capacity required to reach the target. Third, the added capacity needs to be installed in an electricity market which is going through profound changes in terms of liberalisation and creation of the Iberian Electricity Market (MIBEL). Fourth, the cost of the policy is becoming increasingly high; the extra cost of the feed-in tariffs alone will be about €100 per household in 2010.

The ambitious solar water-heating programme has the potential for harnessing this renewable energy which is amply available because of the favourable climate in some parts of Portugal. The programme has some quite innovative elements such as applications for industrial processes. The close collaboration with the manufacturing industry as well as the establishment of a monitoring agency is likely to enhance the effectiveness of the programme.

The major policy implemented to achieve the targets is the feed-in tariff scheme for renewable electricity. The tariffs, which have the same impact as subsidies, are generous and appear to give enough incentives for potential investors based on the applications for licences. This system strongly boosts the domestic production of renewable energy, wind in particular. However, given the progress to date with 500 MW of wind capacity under construction or contracted, additional efforts will be needed so that the required extra capacity will be built in time to meet the target.

Despite the expected increase in renewable electricity generation, there is some room for vigilance in the implementation of the feed-in tariffs. Feed-in tariffs, depending on the design, may not provide strong incentives for cost reduction and it may be the producers, not consumers, who enjoy the benefits of cost reductions unless benefits are passed through owing to competitive pressures. Consequently, the scheme should be carefully monitored and comparisons made with other policy options.

The government should seek to improve the current feed-in tariff mechanism so that final consumers could benefit from the cost reduction while ensuring investor confidence in renewable energy projects. The Portuguese market is not sufficiently large to drive costs, but will, to a certain extent, benefit from cost reductions in the larger market. However, Portugal has increased the level of feed-in tariffs three times over nine years to a level which is, in the case of wind power for example, higher than in some other IEA member countries. In this context, the global cost levels should be monitored very closely; periodic reviews of the feed-in tariff level for new investments are necessary to incorporate the most recent global trend of cost reduction. Indexing the feed-in tariffs with inflation is not justified by the changes in the cost of operating renewables. Periodic reviews and removal of indexing would not only reduce the cost to consumers, but, more importantly, would in time stimulate decreases in generation costs. Furthermore, feed-in tariffs should only apply for a predefined time period for each new project. After the initial investment of renewable energy installations has been fully amortised the existing system unduly punishes other generation options. Lastly, given the establishment of MIBEL, policy co-ordination with Spain is essential. Closer co-operation with Spain can also bring learning benefits for both countries since cost reductions are more difficult to achieve in small markets as they are largely driven by global learning.

Government support for renewables is a sound policy because the positive environmental protection and energy security externalities of renewable energy are not currently captured by the market. When mature, however, renewables need to be exposed to competition from other mature energy sources. The precondition for this will be adequate internalisation of the different externalities. While it is a very difficult task to quantify and internalise the full range of externalities, progress towards this end should be assessed and, if achieved, the subsidies for renewables should be gradually reduced and eventually phased out. As a first step, the level of the feed-in tariff needs to be reviewed when EU-ETS is introduced because the gap between the generation cost of renewable energy and that of conventional power sources will be narrowed by incorporating carbon costs. In assessing the subsidies for renewables, it should be noted that other policies, such as increased investments in energy efficiency, carry similar benefits and can have a lower cost than renewables. There appears to be abundant potential to improve energy efficiency in cost-effective ways, particularly in Portugal (see Chapter 5). In addition, the interaction of the feed-in tariffs and MAPE subsidies should be reviewed. While such a combination of policies is consistent with practices elsewhere, it should be borne in mind that total government support should provide both security to investors and cost-minimisation to consumers. Portugal, like many countries, has established two instruments towards these goals, but they should be constantly monitored and modified, if necessary, to be sure that one goal is not met at the expense of the other.

The lead times for renewable electricity generation projects appear to be too long, creating a risk. In particular, the lengthy procedure for environmental licences could create a challenge for the renewables target as it reduces investor confidence. In addition to avoiding investments in renewables altogether, investors may opt for smaller-scale projects because the cost risks associated with long licensing processes increase. On the other hand, smaller projects can increase consumer costs. Therefore, licensing procedures should be streamlined, for example by introducing environmental impact assessment (EIA) on a regional rather than on a project-by-project basis and by clarifying the criteria applied in the EIAs. Another way to simplify the processes is to introduce a "one-stop-shop" for licence applications.

With significant growth, intermittent renewable energy sources, principally wind power, could have a negative impact on the stability of grid and optimal operation of baseload power capacity. Problems could occur either when wind capacities are providing too little or too much power. Existing wind capacity is small and, therefore, not yet a concern, but this may not be the case when its share of total electricity generation increases further. There are no universal limits above which problems are encountered, so the local conditions in Portugal need to be studied. To an extent, technical solutions and business and regulatory practices can help the integration of large wind capacities. Also, the wide geographical range of good wind sites, and the fact that generation is seasonally high in different regions, suggest that Portugal may have some advantages for incorporating relatively large amounts of intermittent wind. Solutions require research and development, new management techniques and possibly additional investments in networks. However, the continuous modernisation of the grid may reduce the incremental investment attributable to added wind capacity, and hence, the corresponding financial burden on wind power companies.

RECOMMENDATIONS

The government of Portugal should:

- *Review the current feed-in tariff scheme in order to assure cost minimisation to consumers while ensuring investor confidence. Assess the benefits of incorporating incentives for cost reduction through gradually reducing the tariff level and the duration of the buy-back period. Also review the interaction of the feed-in tariffs and subsidies to determine when and which incentives can best be reduced.*

- *Assess progress towards a competitive renewable energy sector with a view to ensuring a stable investment environment until targets are met. Phase out the subsidies in the longer term when the different positive and negative externalities of renewables and other energy forms have been internalised.*

- *Continue efforts to streamline licensing procedures, including the environmental impact assessment, for renewable energy projects.*
- *Investigate the requirements of reliability and stability of the future electricity network, given the indicative goal of connecting large amounts of wind power to the grid.*

ELECTRICITY AND CO-GENERATION

INDUSTRY STRUCTURE

Portugal's electricity sector is quite complex, with a hybrid market structure. However, under the complexity, the dominant role of Electricidade de Portugal (EDP) is very important.

Inside Portugal, EDP is dominant in both generation and distribution of electricity. Its generating arm, EDP Produção, owns 8 GW of capacity and generated about 69% of Portuguese electricity supplied to the grid in 2003. EDP Distribuição supplies 99% of grid electricity to end consumers. EDP also has minority interests in the transmission company, Rede Eléctrica Nacional S.A. (REN), (30%), and in the two large independent power producers Tejo Energia and Turbogás. Its retailing arm, EDP Energia, sells about two-thirds of the consumption in the liberalised segment of the market, with the remainder sold by Spanish-owned companies.

EDP was formerly a state-owned company but has been progressively privatised. While state ownership is limited to around 30% of total, the government has a great influence over EDP. In the past year, the government nominated and appointed its new CEO who had championed a merger between EDP and GALP and has been heavily involved in the merger discussions.

EDP and the other market participants operate in a hybrid market structure which has been in place since 1995. The current market structure consists of a public electricity system (PES) and an independent electricity system (IES), as shown in Figure 20. The PES is mostly characterised by capacity planning and purchasing power agreements between generators and REN. The IES consists of the non-binding system (NBES) which is characterised by free contracts between generators and eligible consumers and the generators in the special regime which includes co-generators and generators from renewable sources.

PUBLIC ELECTRICITY SYSTEM

The public electricity system is a regulated system designed to ensure public service, for example security of supply and obligation of supply. Generation in the public electricity system comprises CPPE (generation owned by EDP Produção) as well as the two independent power producers (IPPs), the 584 MW coal-fired generation plant at Pego, owned by Tejo Energia, and a 990 MW CCGT at Tapada de Outeiro, owned by Turbogás. Together, these plants accounted for 85% of Portuguese electricity generation in 2003. This system also includes the transmission grid owned and operated by REN and the distribution company, EDP Distribuição.

Figure 20

The National Electricity System

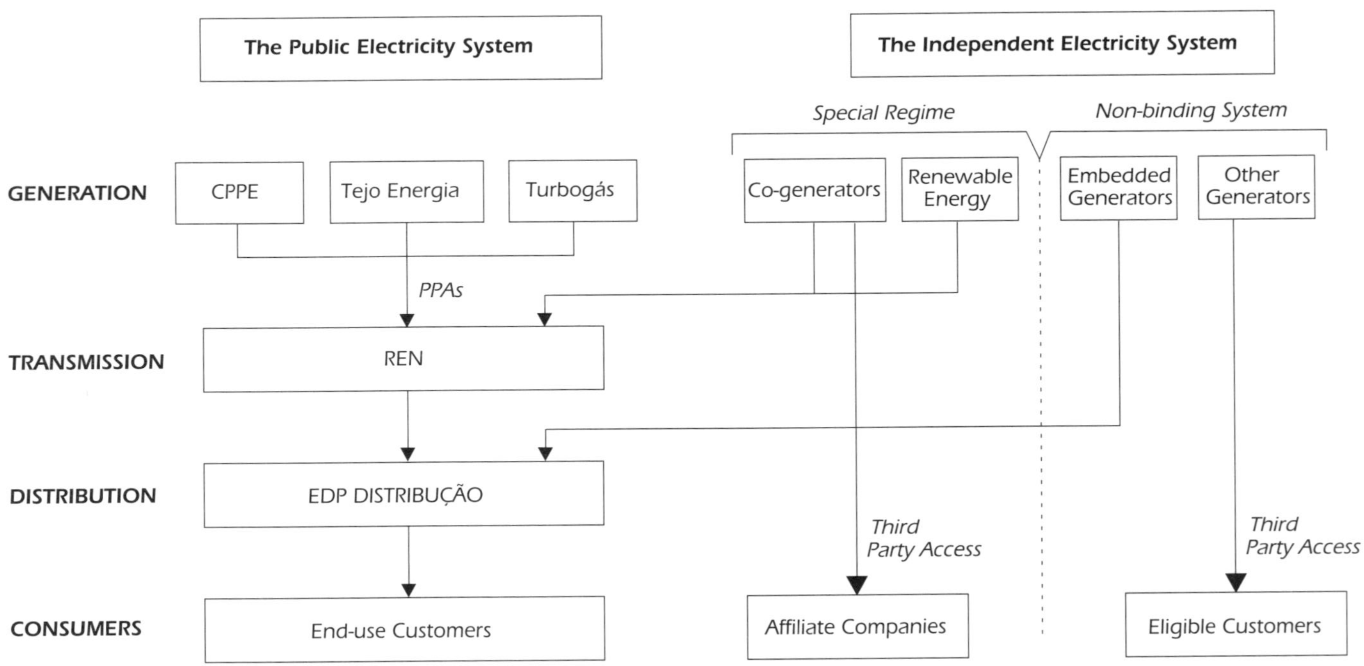

PPAs: Power Purchase Agreements.
Source: EDP and ERSE.

Under this system, the need for additional power plants was determined by a planning process. Expansion plans for new capacity were reviewed by the regulatory authority which gave a non-binding recommendation. New capacity was then awarded by a tendering process.

Generators in the public electricity system sell power to the transmission company, REN, through a long-term power purchase agreement (contract for acquisition of energy – CAE). The CAE, designed to last the economic life of the plant, aims to achieve a regulated return on investment. Most of the revenue for producers comes from capacity charges. Energy charges are designed to allow producers to recover actual fuel and operating costs.

INDEPENDENT ELECTRICITY SYSTEM

The independent electricity system has two components. The first is the "special regime", which includes co-generation and most production other than large hydro renewable production. This sector sells power directly to REN, a tariff having been decided by the government. In 2003, this sector supplied 3.7 TWh, or about 9% of total supplies, to the grid, of which around 1.5 TWh was produced by wind and small hydro.

The other element of the independent sector is the NBES. This includes other independent power production and imports whose power is sold directly to consumers that are able to choose their power supplier (qualifying consumers). This production, excluding imports, amounted to less than 2% of production in 2003.

The regulator has been gradually increasing the number of consumers eligible to buy electricity through the NBES. The share of electricity consumers eligible to choose suppliers has been growing in steps. By 2002 all high- and medium-voltage consumers were able to choose their supplier. This was extended to some smaller consumers (special low-voltage in January 2004). The government intends to extend the choice to all consumers in July 2004.

Eligible customers have the option of remaining under the public electricity system on regulated tariffs or moving to the NBES. The participation rate by consumers in the NBES has varied by customer size. Although high-voltage consumers have been able to change supplier since 1999, only three of the 125 high-voltage consumers had moved from regulated tariffs by the end of 2003. However, over 30% of electricity supplied to medium-voltage consumers in 2003 was purchased through the NBES. Two-thirds of this energy was purchased from the retailing arm of EDP, with Sodesa (owned by Endesa and the Portuguese holding company Sonae) and Iberdrola supplying the rest.

Table 8

Electricity Consumption and Customer Choice by Voltage Level, 2003

	Number of customers	*Date of eligibility*	*2003 PES consumption (GWh)*	*2003 NBES consumption (GWh)*	*Share of NBES consumption*
High-voltage	128	February 1999	4 755	114	2.3%
Medium-voltage	20 897	January 2002	8 600	3 935	31.4%
Special-low voltage	28 111	January 2004	3 050	-	-
Low-voltage	5 718 780	July 2004	18 463	-	-
Total	**5 765 916**		**34 868**	**4 048**	**10.4%**

Source: EDP.

DEMAND, SUPPLY, TRANSMISSION AND TRADE

ELECTRICITY DEMAND

In 2003, final electricity consumption amounted to about 43 TWh. Electricity demand has been growing steadily in Portugal, averaging 4% per annum since 1990, significantly greater than the rate of GDP growth which averaged 2.5% over the same period. Demand growth, while similar to neighbouring Spain, is about double the IEA Europe average. Growth has been strongest in the commercial and residential sectors, with moderate growth in the industrial sector (Figure 21).

Peak demand of 8 GW occurs in winter. A secondary, but much lower peak, is observed in summer. Available capacity is approximately 9.2 GW. Hydropower accounts for approximately 4.1 GW of this capacity but its actual availability can be considerably less owing to rainfall variations.

ELECTRICITY SUPPLY

Electricity is supplied by a combination of coal, hydropower, oil, gas, biomass and wind power. Coal is the largest source supplying about 31% of electricity generation. Hydropower is a substantial source, but output is highly variable, supplying 15 TWh (34%) of power generated in 2003 but only 7.5 TWh (16%) of power generated in 2002. Much of the difference in hydro output is made up by increasing oil-fired generation which accounted for 25% of power generated in 2002 but only 15% in 2003. Gas-fired generation, mainly from new combined-cycle power plants, contributed another 17%. Biomass generation produced about 4% of the power supplied. Other renewables, mainly wind power, were responsible for only 1% of supply. However, this is expected to increase greatly in the coming years thanks to favourable policies for renewable energy other than large hydro (Figure 22).

Figure 21

Final Consumption of Electricity by Sector, 1973 to 2010

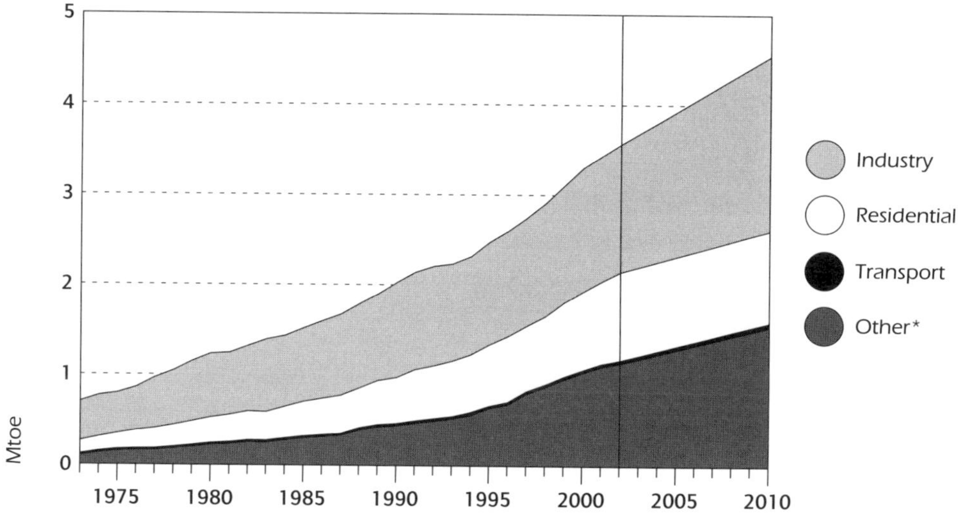

* includes commercial, public service and agricultural sectors.

Sources: *Energy Balances of OECD Countries*, IEA/OECD Paris, 2003; and country submission.

Figure 22

Electricity Generation by Source, 1973 to 2010

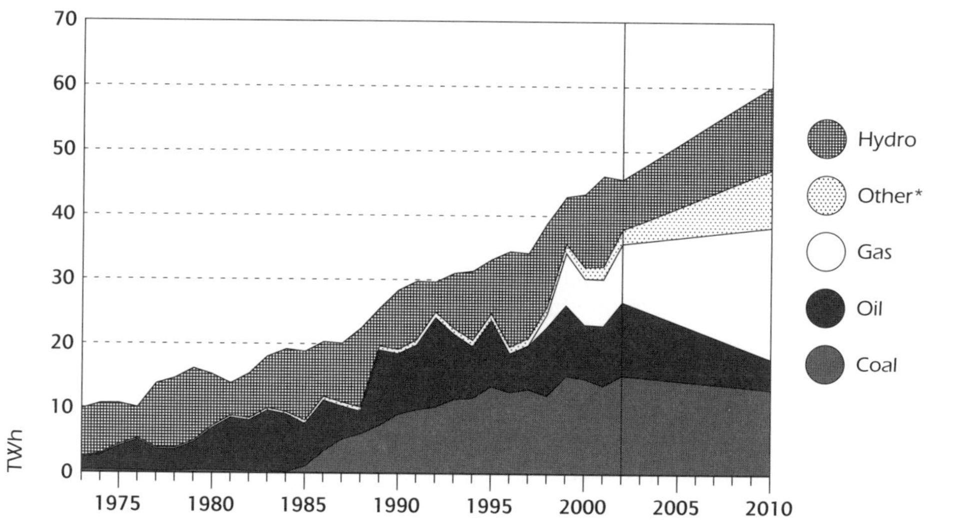

* includes geothermal, solar, wind, combustible renewables and wastes.

Sources: *Energy Balances of OECD Countries*, IEA/OECD Paris, 2003; and country submission.

Significant new power plant construction is planned or under way, much of it by EDP. EDP investments include the TER CCGT power plant (3 x 392 MW), the first unit of which produced power in late 2003. EDP participated in the construction of the Alqueva hydro plant (240 MW) commissioned in 2004 as an engineering company and is also investing in hydropower as well as building the Venda Nova II plant (180 MW) which is to be commissioned in 2004. Two other plants, Baixo Sabor (140 MW) and Picote II (236 MW), are planned to be built by 2010. Among independent producers, Iberdrola has indicated it is studying adding a CCGT plant in Portugal at Figueira da Foz, and Endesa also intends to build two CCGT groups at Pego. There is substantial independent investment in wind power, although here too, EDP will add 580 MW by 2008.

ELECTRICITY GRID

The Portuguese transmission system is owned and operated by REN. At the end of 2003, it consisted of 1 403 km of 400 kV, 2 704 km of 220 kV and 2 434 km of 150 kV power lines and 47 substations as well as the two system control centres near Lisbon and Porto. As the transmission system operator (TSO), REN manages the scheduling of generation and imports and exports to match demand. REN will continue to act as TSO for Portugal in the integrated Iberian market as well as managing the Iberian forward market (OMIP). The government announced the merger of REN with the gas network operator Transgás.

REN is undertaking an ambitious reinforcement programme of the 220 kV network in order to reduce transmission congestion within Portugal and to increase cross-border capacity. Accordingly, new interconnections will be constructed: a new 400 kV link (Alqueva-Balboa) is scheduled to open in December 2004 and a new 400/220 kV substation (near Douro border with Spain, an initial operation only with 220 kV) in 2007, and the reconstruction of the two 220 kV lines (Douro Internacional-Aldeadavila) is scheduled for 2007. These reinforcement programmes will increase cross-border capacity to 1 500 MW by 2008, representing 18% of the anticipated Portuguese peak demand. Major projects are summarised in Table 9.

According to REN, substantial additional investments will also be needed to reinforce the network within Portugal to accommodate new renewable electricity generation, particularly from wind which is expected to account for 21% of installed generating capacity by 2010. In addition, REN anticipates that extra generating capacity will be needed for backup and reserve margin for the security of system operation.

The distribution network is operated by EDP Distribuição under a licence granted by the government. EDP operates under concession agreements with municipalities, pays annual concession fees to these municipalities and recovers costs from low-voltage consumers. The main data for the distribution network are set out in Table 10.

Table 9

Current Major Transmission Network Expansion

Project	*Type*	*Commissioning*	*Impact*
Alqueva-Balboa	2 x 75 km 400 kV line	2004	Increase international transmission capacity (new line)
Alto-Lindoso-Cartelle	400 kV line	2004	Increase international transmission capacity (expansion)
Douro Internacional -Aldeadávila	220 kV line	2007	Increase international transmission capacity
Pereiros -Zêzere-Carregado Santarém	2 x 153 km 220 kV line	2004	Reduce internal constraints and facilitate international flows
Valdigem-Bodiosa Bodiosa-Paraímo	2 x 60 km 400 kV line	2005/2006	Increase north-south flows and facilitate imports
Sines-Portimão	1 x 97 km 400 kV line	2005	Increase reliability/ decrease congestion
Pego-Batalha	2 x 81 km 400 kV line	2005	Increase reliability and support for 400 kV import line
Pedralva, Anadia, Falagueira, Douro Internacional	New substations with phase shifters	2005-2006-2008	Better use of 400 kV system, increase reliable import capability

Sources: REN and UCTE.

Table 10

EDP Distribution Lines, 2002

(km)

Distribution lines	*Overhead*	*Underground*	*Total*
High-voltage (60/130 kV)	7 097	357	7 454
Medium-voltage (< 6/10/15/30/40 kV)	51 460	10 995	62 455
Low-voltage (1 kV)	96 265	23 973	120 238
Total	**154 821**	**35 325**	**190 147**

Source: EDP.

Distribution losses are relatively high in Portugal at 7.5% of total electricity delivered, compared to an IEA Europe average of around 5%. Similarly, according to an analysis carried out for the Council of European Energy Regulators, the average interruption time in the distribution network is at least double that of eight other European countries in their survey. Reliability problems are acute in rural areas. Regulations established by the DGGE in 2003 aim to encourage increased investment in distribution by providing compensation payments to consumers for interruptions.

As a consequence, significant investment is being made to improve the quality of electricity services. Distribution grid investment in 2003 amounted to €266 million and will continue at similar levels through 2006. It is hoped this will reduce interruption times by over 40%, bringing values closer to European averages. This investment is occurring while costs are being reduced through a restructuring programme aimed at reducing the workforce by 20% by 2007.

ELECTRICITY TRADE

Electricity trade between Spain and Portugal has normally been quite limited, in part because of the level of interconnection capacity between the two countries. The opening of the electricity market in Portugal has allowed Spanish producers to sell to Portuguese consumers. However, only 400 MW has been available for trading purposes. Trade has increased markedly as imported power supplies most of the consumers in the liberalised market who are not supplied by EDP (Table 8). The introduction of the Iberian market and the additional investments in cross-border capacity is expected to lead to much higher levels of electricity trade (Table 11).

Table 11

Electricity Trade, 1999 to 2003 (GWh)

	1999	*2000*	*2001*	*2002[1]*	*2003[1]*
Import	626	1 762	1 307	2 185	3 078
Export	1 483	845	1 068	286	284
Net import	**-857**	**917**	**239**	**1 899**	**2 794**

[1] The consideration of simultaneous exports, instead of these hourly balances of the interconnection, suggests larger activity. In 2003, simultaneous imports were 4 565 GWh and exports 1 771 GWh. In 2002, the corresponding figures were 2 750 GWh and 851 GWh.

Source: REN.

Table 12

Planned Increase in Net Transfer Capacity between Portugal and Spain (MW)

	2003	*2004*	*2005*	*2006*	*2007*	*2008*
Exports	700	700	1 260	1 440	1 620	1 890
Imports	850	850	900	1 080	1 260	1 530

Source: REN.

PRICES AND TARIFFS

Retail electricity prices for Portugal are higher than average in the IEA for household consumers (Figure 23). Interruptible rates are offered to large industrial customers. According to Eurostat statistics[4], which are more disaggregated than IEA data, industrial prices in Portugal are 10% to 20% higher for comparable consumers in Spain (Figure 25).

Figure 23

Electricity Prices for Households in IEA Countries, 2002

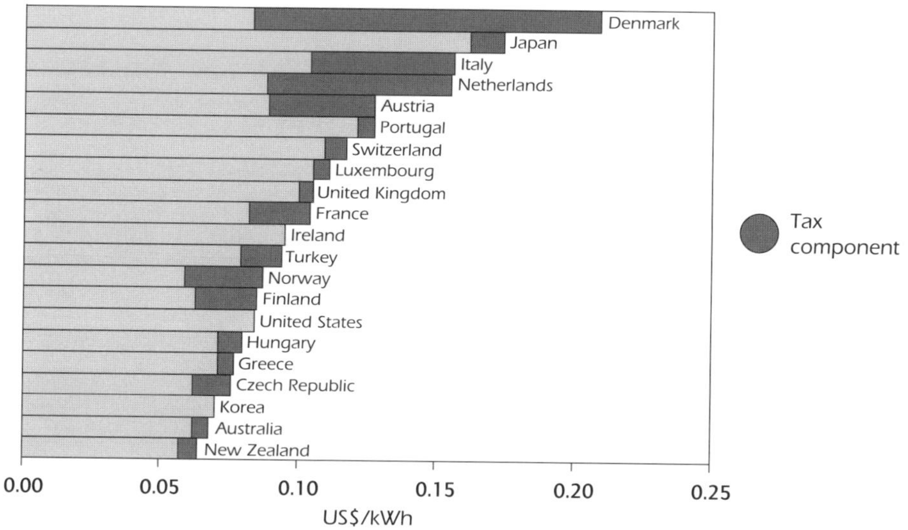

Note: Price excluding tax for the United States. Tax information not available for Korea. Data not available for Belgium, Canada, Germany, Spain and Sweden.

Source: *Energy Prices and Taxes*, IEA/OECD Paris, 2003.

Wholesale electricity prices in the binding system amounted to €49/MWh in 2003. These are significantly greater than those of Spain (€37/MWh in 2003) and thus there is a significant potential for a decrease in wholesale market prices in Portugal once the Iberian market begins operation[5].

4. Portugal's reported average industrial prices are based on an average industrial customer and thus differs from the measure used by most other countries. Therefore, Eurostat data comparisons, based on average prices for different types of industrial customers, are used here.
5. Although these gains may be offset by a charge to recover stranded costs.

Figure 24

Electricity Prices for Households in Portugal and in Other Selected IEA Countries, 1980 to 2003

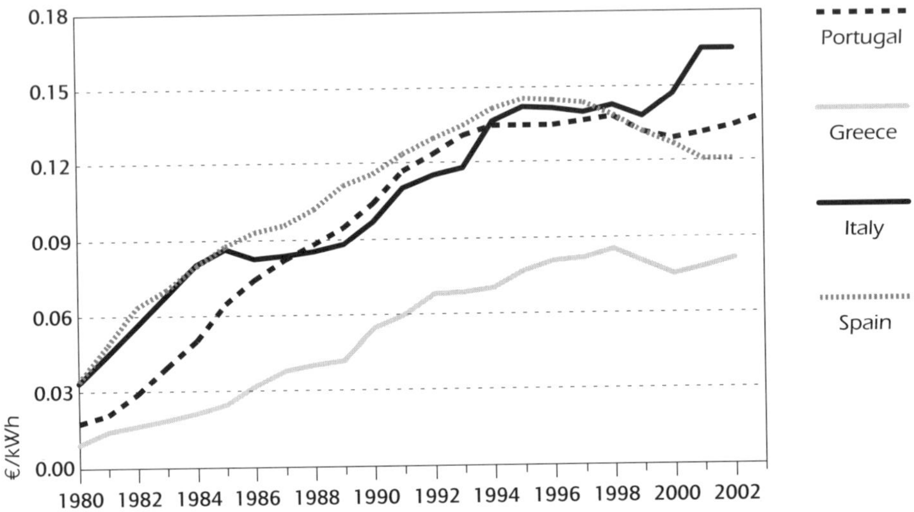

Source: *Energy Prices and Taxes*, IEA/OECD Paris, 2003.

Figure 25

Electricity Prices for Industry in Portugal and Spain, 1 July 2003

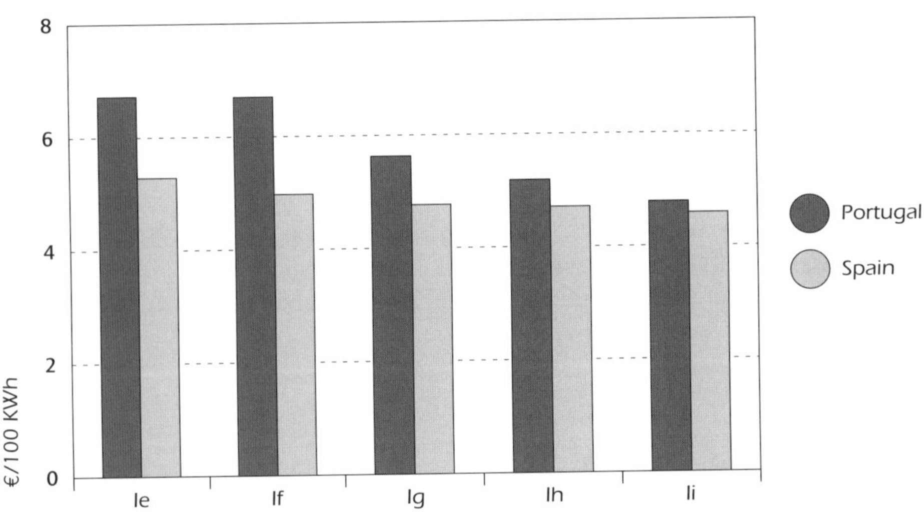

Source: Eurostat.

Customers in the binding system are subject to regulated retail electricity tariffs. In the period 2001-2003, tariffs have increased at an average 2% per year, somewhat less than the rate of inflation.

Electricity network prices are regulated by ERSE in the binding sector through a price cap mechanism with a three-year period. The price caps establish changes in distribution charges at the rate of inflation, less an "X factor" to account for improvements in efficiency by the distributor. The "X factor" was set at 7.5% for high- and medium-voltage and 6.6% for low-voltage from 2002 to 2004 for distribution use of system tariffs, which is a very challenging target.

MARKET REFORM

Although EDP plays a dominant role within the Portuguese market, it is still smaller than the two largest generators in Spain (Table 13). Spanish generators were already taking advantage of existing interconnections between Spain and Portugal to sell power to eligible consumers in Portugal. Iberdrola has a minor interest in the Portuguese gas sector and is studying the development of a new CCGT plant in Portugal. EDP in turn had purchased 40% of the Spanish utility Hidrocantábrico in 2002 and, until the autumn of 2003, had a 3% shareholding in Iberdrola.

Table 13

Iberian Electricity Market Shares of Largest Utilities, 2002

(%)

	Generation	*Distribution*	*Supply*
Endesa (Spain)	38	32	37
Iberdrola (Spain)	25	35	38
EDP (Portugal) + Hidrocantábrico (Spain)	19	19	8
Union Fenosa (Spain)	11	12	13
Other	7	2	5

Source: Iberdrola.

It, therefore, seems logical for the Portuguese and Spanish electricity markets (see box for a description of the Spanish market) to be combined into a single electricity market in order to create greater competition among power producers and a greater choice for consumers. In November 2001, the governments of Portugal and Spain signed a protocol to create an Iberian electricity market (MIBEL) with the goal of starting a new trading structure by 2003.

The Electricity Market in Spain

The operation of the Iberian market will closely resemble that of the Spanish market. This box provides a brief description of the day-ahead market in Spain.

The electricity market in Spain is based on a legal framework which sets out the commercial agreements that govern the sale of power in the market. The market operator (OMEL) does not buy or sell electricity. Generators and buyers trading in the pool must record their data in the Administrative Registers and confirm their adherence to the market rules.

Market operation is based on a day-ahead market. The suppliers and buyers submit bids consisting of two components: the amount of electricity to be traded and the proposed selling or purchasing price. OMEL matches these bids and the resulting marginal price is equal to the last production block which is needed to meet the demand. All generators, including those who offer electricity under the marginal price, will receive the same remuneration based on the marginal price.

The market operator modifies the accepted bids in consultation with the transmission network manager to relieve technical restrictions, and then runs a secondary market for backup power for the day. In addition, there is an intra-day market with up to six sessions, which allows further adjustments to the schedule by producers and customers to smooth supply by individual generators and to deal with events on the day such as generator unavailability. Ancillary services are acquired by bidding and auction mechanisms managed by the system operator.

The main participants in the market are the utilities, the qualified customers, agents authorised to trade electricity and importers of electricity. The utilities play several roles, acting as generators, distributors, suppliers purchasing power for customers remaining on regulated tariff, and agents acting for liberalised customers with whom they have a contract. Qualified customers can opt to enter as market participants but can also remain with a regulated supplier on a regulated tariff. All market participants must register with the Ministry of Economy after demonstrating they meet the criteria for qualification. For customers, the only material criterion is that they purchase enough electricity to qualify.

Generators over 50 MW must be registered with the market and will be dispatched through the market operators. Customers can then purchase power directly from the market through a contract with a retailer, or through a physical bilateral contract with a supplier.

Work on the creation of the market has involved governments, regulators and system operators in both countries. The original schedule proved to be overly ambitious, given all the changes to legal statutes, market definition and regulation required. Harmonisation of technical planning and operational issues, and resolution of technical aspects relating to increasing interconnection capacity, also had to be achieved.

The main elements of the MIBEL include:

- Bilateral trading: this is expected to be the main form of electricity trading in the Iberian market. All ordinary power producers must make their capacity available to the market. Furthermore, producers will be obliged to publish, on a consolidated basis, the prices under these contracts. Contract term length will be limited to two years in order to build up electricity market liquidity. Restrictions are also to be placed on vertical sales, *i.e.* between the sale of power from one producer to its retailing arm.
- A day-ahead market: this hourly market will be operated by the Spanish market operator based on the market currently operated in Spain.
- Physical forward markets, with standardised contracts, are to be developed and operated by the Portuguese TSO, REN.
- Balancing markets: each system operator will operate a balancing market.
- Ancillary services markets: separate markets for Spain and Portugal will be kept in early stages of the operation of the MIBEL because of limitations in interconnections. It is possible that these will be replaced by a single market should interconnections prove sufficient.
- Capacity guarantee: each system operator will operate a balancing market protocol mapping out a tight schedule of changes to meet the requirements of the new commercial and technical model of operation.
- Congestion management: in light of the expected congestion on the interconnections between Spain and Portugal, use of capacity auction or a market splitting mechanism is envisaged.

In January 2004, the governments of Portugal and Spain signed an agreement to establish a single Iberian market to open on 20 April 2004. Nevertheless, several important details remained to be defined to create a common market and the market opening date has been delayed. Many issues have to be resolved before the start of operations of the MIBEL including:

- The mechanism for managing congestion, including the mechanism for allocation of capacity and cost recovery.
- Stranded cost recovery mechanism.
- Harmonisation of emissions trading regimes in the two countries.
- Harmonisation of approaches on "special regime generation".

The regulators in Spain and Portugal have voiced concerns about the concentration in the Iberian electricity market of the three largest generators. They support the development of virtual power plant auctions as a mechanism to decrease concerns about market power within Spain and Portugal.

STRANDED COSTS

Most of the electricity in Portugal has been tied to the public electricity system under the contracts for the acquisition of energy (CAEs). Nearly all these contracts last until at least 2010, over 8 GW of capacity. This contract level drops to 4 GW of capacity by 2018, with the final contracts expiring in 2025. A prerequisite for creating competition in the supply of energy within Portugal for the Iberian market is to allow Spanish producers to compete with Portuguese producers to supply either regulated consumers or consumers in the market. Therefore it has been recognised that CAEs need to be reduced and ideally terminated to maximise competitive supply options. In fact, one of the conditions for the MIBEL to open is that CAEs must account for less than 50% of total electricity sales in Portugal. Since wholesale market prices in Spain are lower than those in the CAE contracts, the expectation is that power producers would lose money if they were forced to terminate their CAEs. As a consequence, there will be stranded costs or "costs of transition to competition" (CTCs) for which the power producers will expect to be compensated.

Spain already has a CTC mechanism which compensates producers on the basis of an expected market price and recovers these costs through an additional charge to consumers. The Spanish law specifies a total amount to be compensated. The annual compensation paid is adjusted to the market price and is also capped at a particular price level. All CTCs have to be recovered no later than 31 December 2010.

In their evaluation of the organisational model for the MIBEL, the regulators in Spain and Portugal noted that it was certainly sensible to have a mechanism for Portuguese power producers compatible with that of Spain but also suggested that adjustments take place over a longer period in order to reduce incentives for adjusting short-term operating strategies.

They also recognised that simply ending the CAEs would not be sufficient in itself to eliminate concerns about market power in Portugal. Recently, both regulators have raised the possibility of pursuing a different CTC mechanism that would both fairly compensate the power producers for the value of their contracts and have the benefit of creating greater competition among generators. This mechanism would involve auctioning the power production rights to the contracts, currently owned by REN, to third parties. Power producers would continue to receive revenue according to the contracts, but the owners of the power produced would be able to resell at actual market prices. The sale of "virtual power plants" would have the benefit of increasing the number of participants selling power within Portugal and hence increase competition in the Portuguese part of the electricity market. The regulators advocated that the same process be applied to power plants owned by Endesa and Iberdrola as well, because they consider that failure to do so would only increase Endesa's and Iberdrola's dominance at the Iberian level and might reduce EDP's capability to change the current Spanish market dynamics.

The government has apparently reached an agreement with EDP on the compensation to be paid for releasing EDP capacity to the market. This agreement will release sufficient capacity to meet the obligations under the agreement to open the MIBEL market. Details of the stranded cost compensation mechanism have not yet been disclosed.

CO-GENERATION

Co-generated electricity, principally in the industrial sector, accounted for 1 300 MW of capacity and produced 4.5 TWh of electricity in 2003. Over half the power production is fuelled by oil, the remainder split nearly equally between natural gas and biomass.

Of this amount, about 2.2 TWh was sold by co-generators to the national grid company at a tariff determined by the government. The tariffs paid to the producers are a function of performance of the undertaking and of criteria for availability; an environment premium is added if thermal efficiency of the plant is at least equal to the most efficient in each category. In 2002, the average tariff paid to co-generators amounted to €66/MWh produced.

The government expects co-generation to increase considerably to 2 000 MW by 2010. In principle, there is sufficient market to achieve this increase in capacity. However, in practice there remain some barriers to this increase. It is becoming increasingly difficult to find suitable large consumers of heat in the industrial sector because of the reductions in the industrial sector. Furthermore, existing industrial customers are less willing than before to commit to a long-term heat contract which is necessary to manage the risks associated with it.

The commercial sector represents a potential growth opportunity and the government is taking steps to inform commercial consumers of the opportunities to use co-generation. However, with very limited needs for heat, it is more difficult to make a good economic case and market penetration has been very limited.

CRITIQUE

The electricity market in Portugal has developed significantly since the last in-depth review but remains dominated by EDP. A significant portion of medium-voltage customers have moved from the regulated system to the market. Yet the majority of these have simply changed to the unregulated arm of EDP, the dominant generator, to take advantage of more flexible rates. The largest consumers remain on regulated tariffs, thanks to prices lower than those available from the market.

Large Spanish power companies are also participating in the market as suppliers and have expressed interest in developing power plants inside Portugal. However, EDP remains the dominant supplier and its subsidiaries are building the majority of the new capacity.

There are several initiatives to increase electricity market competition within Portugal. The increase in physical interconnections with Spain would increase competition within the Portuguese market and improve security of supply. The development of the natural gas network has increased the feasibility of independent entry into power generation within Portugal. Even the policy to greatly increase renewables will increase the number of independent suppliers in Portugal, although here too, EDP intends to develop a significant proportion of the total new renewable power generation.

The political initiative to create a common Iberian market would be an important step in creating a competitive market. But the progress towards this goal appears to be behind schedule. A number of technical and regulatory questions related to managing transmission congestion, and harmonising approaches to "special regime" generation and to emissions trading allocation, will need to be dealt with. The creation of an Iberian market requires suitable legislation in both countries, yet the detailed legislative proposals have yet to be published. Major market players, regulatory authorities, and most importantly the market operators in Spain and Portugal, need the detailed proposals to finalise their preparations for this market.

However, without further measures, it is not likely that electricity supply in Portugal will be truly competitive for years. Enhanced transmission links, while very helpful in increasing competition in generation, could easily become congested. Price differences between Spain and Portugal are likely to persist with EDP as the dominant producer.

In other countries concerned about dominant suppliers, the strategy that has been employed is to put some generation of the dominant supplier under contract and so remove incentives to push up market prices. Such a contractual agreement is currently under consideration.

However, the need to unwind the long-term power purchase agreements between REN and EDP could be used to enhance competition in the Portuguese market. One possible way would be auctioning the contractual obligations by REN to new suppliers, which would make more electricity available to new entrants. These could then compete to sell it to the liberalised consumers for the duration of the current CAEs. This would have the impact of reducing EDP's market share until better interconnections and new investments by independent generators can be achieved.

However, the question of concentrated generation in the Iberian electricity market can only really be solved if addressed at the Iberian level. EDP is still smaller than the two largest Spanish generators and there is a reluctance to diminish EDP's size while large Spanish producers are unaffected. In this situation, it is still possible to enhance competition in generation in the Iberian market through the use of "virtual power plant" auctions of capacity as is used in France. The regulators of Portugal and Spain have proposed such auctions of parts of the capacity of the three largest market players (Endesa, Iberdrola and EDP) to increase competition in the Iberian market. This proposal needs careful consideration by both governments. Another possibility would be to require an exchange of generating assets between EDP and the Spanish producers. This would increase competition within Portugal. However, it would still leave the three largest producers in a dominant position in the Iberian market.

In any event, close monitoring of the prices and bidding behaviour of the major players in the Iberian market, particularly in terms of wholesale prices in Portugal, will be necessary. This will require involvement of the appropriate competition authorities. This is vitally important during the early stages of the Iberian market when only limited interconnection capacity exists between the two countries. Even as this capacity is expanded, however, the likelihood of congestion on these lines will limit competition to supply electricity within Portugal. Effective management of access to these lines will be important to maximise possibilities for competition to the benefit of Portuguese consumers.

Similarly, while there is significant political will to open the electricity market to all consumers in Portugal on 1 July 2004 (as is already the case in Spain), it is still uncertain that meaningful competition to serve these customers will develop quickly. Legislation and regulations will be needed to clarify the rules for small consumers to switch suppliers.

Setting regulated tariffs for the largest consumers below market prices means that few of these consumers are willing to purchase electricity on the market. This is counter-intuitive to the experience in liberalised markets where the largest consumers are generally the most active in selecting new suppliers. Regulated tariffs are expected to remain in place until at least 2007. Large consumers will resist removal of these tariffs, particularly when large consumers in Spain also receive highly favourable discounted electricity rates. The solution must be resolved at the Iberian level with the elimination of all regulated electricity rates beginning with the largest consumers in both countries.

A well functioning electricity and natural gas market can enhance the security of supply. A policy of capacity payments has not been proven to be an effective or efficient policy to secure adequate supply. Most other governments that have looked at capacity payment options have rejected them as an inefficient and possibly ineffective policy. Thus, if the capacity payment policy is adopted in the Iberian market, there should be a review of this policy by both governments. This review should include a consideration whether the payment should be either removed or whether, as a temporary measure, it should be replaced with a more efficient instrument. Enhancing the demand response of consumers to the market price of electricity will also increase the efficiency of the market and reduce price volatility.

Above all, the launch of the MIBEL will require even closer co-operation between Spain and Portugal to deal with longer-term issues that will affect the Iberian electricity market. The longer-term issues include developing common approaches to the security of electricity supply, allocation of carbon emissions permits and the rules applied for special regime generation.

The addition of substantial new generating capacity will entail significant investment in transmission grid reinforcement. The requirement for power producers to participate in these investments is a good practice. There will also be a need to acquire significant additional backup capacity to deal with the variable output of wind generators. This will further add to the cost of electricity.

Co-generation in Portugal has developed greatly thanks to highly favourable feed-in tariffs established by the government. The government proposes to increase the share of co-generation by 2010 to account for 18% of energy consumption. Although industry sources consider a 500 MW increase in capacity feasible, an extra 700 MW is needed, making achieving this target challenging. There are limited new opportunities to develop co-generation because of a lack of suitable heat customers, which entails a high cost of additional co-generation development. Even where heat customers can be found, economic conditions make it difficult for them to commit to a long-term heat contract. The uncertainty over CO_2 charges is a further deterrent.

Given concerns about the size of the market and the cost-effectiveness of the additional co-generation, this target should be reconsidered. There are, however, significant opportunities to improve the environmental performance of some existing plants through the introduction of newer equipment and through switching fuel from oil to natural gas.

RECOMMENDATIONS

The government of Portugal should:

- *Finalise the new legislation to carry out the objective of creating an Iberian market and to open the market to all consumers.*
- *Ensure the development of adequate interconnection capacity and its fair allocation.*

- *Consider at the Iberian level a mechanism to increase competition in generation through the use of "virtual power plant auctions" to make the energy available to suppliers other than major Iberian players.*
- *Monitor price developments in the Iberian electricity market, and be prepared and able to act promptly should concerns about manipulation of electricity prices arise.*
- *Develop a timetable for the phase-out of regulated power supply tariffs at the Iberian level and consistent with the 2003 EU Electricity Directive.*
- *Monitor generation investment developments in the Iberian electricity market. Consider removing the capacity payment or, as a temporary measure, replacing it with a more efficient instrument.*
- *Encourage the development of demand response mechanisms to enhance the security of supply and decrease the volatility of electricity prices.*
- *Re-evaluate the feasibility of the co-generation target to ensure that it is based on useful heat demand. Improve the environmental performance of some co-generation projects by encouraging the use of natural gas instead of oil.*

RESEARCH AND DEVELOPMENT

GENERAL ENERGY R&D POLICY

The Ministry of Economy in charge of energy policy does not have an energy R&D strategy or programme. In the national science, research, development and technology policies driven by the Ministry of Science and Higher Education, energy R&D is not regarded as a priority. However, energy is a core area of R&D activity of the National Institute for Engineering and Industrial Technology (INETI) which depends on the Ministry of Economy. INETI proposes projects for public funding based on its own evaluation of the priority needs of government energy policy. The proposals also heavily depend on the interest and competence of researchers and the availability of funding. As a consequence of this process, INETI has identified the following objectives for non-nuclear energy R&D:

- Structural objectives:
  - i. Use of endogenous resources and diversification of energy sources.
  - ii. Energy efficiency in different economic sectors.
  - iii. Energy in environmental treatment processes.
  - iv. New energy vectors.

- Horizontal objectives:
  - v. Implementation of measures to respond to international commitments.
  - vi. Energy technology deployment to modernise the national economy.

THE R&D BUDGET

Portugal's public energy R&D budget is very small. The budget was higher at the beginning of the 1990s but was cut rapidly until the mid-1990s (see Table 14). Since then, it has varied between €1.4 and €2.7 million per year (excluding personnel expenses). Budgets for nuclear fission were reported until 1998 and nuclear fusion activities emerged in 2001. In 2002, 66% of the total energy R&D budget was used for nuclear fusion, 21% for fossil fuels, 8% for renewables, 4% for energy conservation and 1% for other areas. There is, however, a lot of uncertainty in the exact breakdown as the statistics are not collected systematically.

Figure 26

IEA Government Budgets on Energy R&D per GDP, 2002

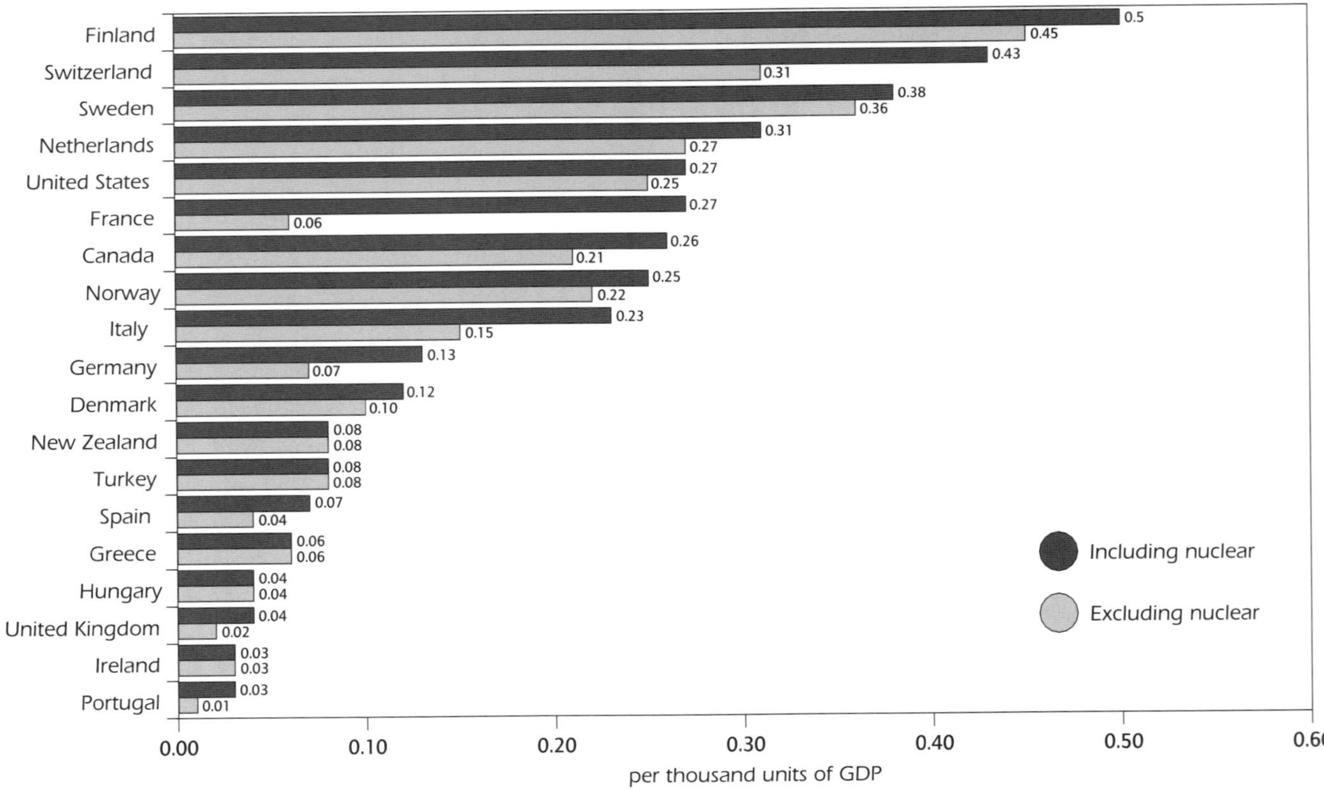

Note: Data not available for Australia, Austria, Belgium, the Czech Republic, Japan and Korea. Luxembourg has no energy R&D programme.

Sources: *OECD Economic Outlook No.74*, OECD Paris, 2003; and country submissions.

Table 14

Government's Energy R&D Budget[1] (Million euros)

Area	1990	1995	1998	1999	2000	2001	2002	2003e	2004e
Energy conservation	1.44	0.67	0.10	0.18	0.20	0.32	0.10	0.09	0.02
Fossil fuels	0.62	0.14	0.18	0.33	0.32	0.17	0.54	0.25	2.46
Renewable energy	1.03	0.45	1.14	1.26	0.76	0.40	0.21	0.24	0.32
Nuclear fission	2.88	0.09	0	0	0	0	0	0	0
Nuclear fusion						1.70	1.72	2.00	2.20
Power & storage	0.03	0.01	0.03	0.02	0.02	0.06	0.01	0	0
Other	1.12	0.02	0.16	0.21	0.17	0.02	0.02	0.02	0
Total	**7.12**	**1.36**	**1.61**	**1.99**	**1.48**	**2.67**	**2.60**	**2.60**	**5.00**

e: estimates.

[1] The numbers presented do not include personnel expenditure.

Sources: *OECD Economic Outlook No 74*, OECD Paris, 2003; and country submission.

The Portuguese energy R&D budget per thousand units of GDP was 0.026 in 2002, which is the lowest among IEA member countries (see Figure 26). There is a commitment amongst EU member States to raise their annual overall R&D expenditure to 3% of GDP. It has not been clarified by the Portuguese government what the role and priority of energy R&D will be in striving for this goal.

R&D ACTIVITIES

In energy efficiency, the activities are mainly in two areas, namely, buildings and optimisation of the efficiency of energy conversion processes. The former involves R&D support to the National Programme for Energy Efficiency in Buildings (see Chapter 5). Examples of the latter are projects for the optimisation of coal blends to improve combustion efficiency and neural modelling and control of coal-fired boiler plants.

In the area of renewable energies and wastes, activities include:

- Photovoltaic energy: photovoltaic and hybrid systems for stand-alone and micro grid applications, grid-connected systems, system modelling and building integration.
- Thermal solar applications: domestic hot water, heating and cooling applications, solar collector testing procedures, thermal electric applications and solar thermal energy storage.
- Passive solar applications in buildings.
- Wind energy: resource assessment and integration of wind power to networks.

- Biomass and waste: development of biogas technologies, biodiesel and bioethanol, waste-to-energy technologies, co-firing of biomass and waste with fossil fuels and biomass quality assurance systems.
- Ocean energy: resource assessment, development of new conversion technologies and grid integration.
- Geothermal energy: feasibility studies of low enthalpy geothermal energy sources.

In fossil fuels the major emphasis is on co-firing with waste and biomass and clean coal technologies, particularly coal/waste gasification and fluidised bed boilers. However, there are other areas of activity too, such as formation/destruction of NO_x in gas flames and petroleum-based waste pyrolysis.

INETI, the Instituto Superior Técnico (IST) and private companies are involved in several demonstration projects in hydrogen and fuel cell technology. These include development of power sources using fuel cells, hydrogen production from chemical hybrids, hydrogen production from biomass gasification and electrolysis with photovoltaics, wind and ocean energy, membrane fuel cells for transport use and development of solutions for decentralised electricity generation.

There are no activities in the area of nuclear fission but Portugal participates in EU programmes for nuclear fusion. The motivation for participating in fusion research is training and providing opportunities for the industry in the long term.

PUBLIC ENERGY RESEARCH INSTITUTES

Nuclear fusion activities are undertaken by the Nuclear Fusion Centre, Associated Laboratory of the Technical University of Lisbon. Public non-nuclear energy R&D is mostly undertaken by INETI but in close co-operation with universities. Other institutes conduct R&D in nuclear fusion and semiconductors for solar photovoltaic. INETI reports to the Ministry of Economy and the Ministry of Science and Higher Education. After restructuring in 2000, energy R&D has been conducted in INETI in two departments, namely Renewable Energies and Energy Engineering and Environmental Control.

INETI gets most of its financing from the government budget and the EU programmes. However, for the last five years, INETI has been improving its co-operation with the industrial sector. Consequently, some of INETI's funds come from the private sector, for example for wind power projects. There are also examples of co-operation with the private sector in technology deployment in waste-fuelled fluidised bed installations and renewable energy technology, and recently interest has emerged in energy efficiency in buildings, biogas and fuel cells.

INTERNATIONAL ACTIVITIES

Portugal places emphasis on the importance of international collaboration given its own limited R&D budget. EU programmes are an important source of financing and Portugal actively participates in them. Many of the projects described above in section on R&D Activities are such co-operation projects.

Portugal participates in seven Implementing Agreements (IEA Framework for International Technology Co-operation), namely Energy Conservation in buildings and Community Systems Programme (ECBCS), Fluidised Bed Conversion, Ocean Energy Systems, Photovoltaic Power Systems, Process Integration, Solar Heating and Cooling, and Wind Turbine Systems. It has also been invited to participate in the Energy Technology Data Exchange. INETI represents Portugal in the IEA Committee on Energy Research and Technology (CERT), in the Fossil Fuels Working Party (FFWP), Renewable Energy Working Party (REWP) and End Use Working Party (EUWP).

MONITORING, EVALUATION AND DISSEMINATION

The Ministry of Economy recognises that there is a lack of coherent activity in the area of monitoring and evaluation. Projects with EU financing are monitored following EU guidelines. The different ministries involved also conduct some evaluation. However, it is recognised that the evaluation results are not applied systematically in planning new activities.

In most cases, the results of energy R&D are disseminated via EU channels since projects receive EU financing. This always involves presenting a dissemination plan.

CRITIQUE

INETI is taking a commendable initiative in identifying activities which support government energy policy. Judging from the breakdown of non-nuclear energy R&D budgets, it has managed to balance the activities among the key policy areas and contributes to important government programmes such as the Solar Hot Water for Portugal programme (see Chapter 7), but it is a concern that expenditure on energy conservation has diminished. It is also positive that INETI is actively seeking possibilities to co-operate with universities and the private sector and to participate in international research programmes. Portugal can greatly benefit from such co-operation given the very small level of public funding. For the same reason, it is positive that Portugal actively co-operates internationally within the EU and IEA frameworks.

However, it is unusual that the ministry in charge of energy policy does not have an energy R&D strategy or programme to help achieve its energy policy goals. The Ministry of Science and Higher Education, which is responsible for national R&D policy, does not regard energy R&D as a priority. Partly because of this lack of interest of the relevant ministries and the absence of a national energy R&D strategy, public funding for energy R&D in Portugal is the smallest of IEA countries with regard to GDP. Furthermore, there are data collection problems on government's energy R&D expenditure.

There are significant energy and environment policy challenges, including meeting the Kyoto target under the trend of declining energy efficiency and connecting large amounts of renewable energies to the networks. The government needs to explore all possible means to respond to these challenges. The possibilities provided by effective energy R&D policy should be fully explored. To implement such a policy, a coherent energy R&D strategy with adequate financing is necessary. To increase the cost-effectiveness of energy R&D and to make better use of the results, it is necessary to develop better monitoring and assessment mechanisms.

Energy issues are becoming increasingly linked with other policy areas such as environment and transport. Similarly, R&D in these issues is becoming more multidisciplinary and there are many links to basic sciences as well. Therefore, close co-operation between the different ministries and the research laboratories under them is necessary. At present, there are some overlaps in the activities of the different laboratories. Another concern is the lack of involvement of both the Ministry of Economy and the Ministry of Science and Higher Education in energy R&D. There is also lack of co-ordination with ministries involved in energy-related R&D, such as the Ministry for Cities, Spatial Planning and the Environment.

RECOMMENDATIONS

The government of Portugal should:

- *Develop a coherent energy R&D strategy with adequate financing to support energy policy objectives.*
- *Better monitor and assess the R&D projects.*
- *Improve the co-ordination between the different ministries involved in energy and related R&D.*
- *Improve the collection of data on governmental R&D funding.*

ENERGY BALANCES AND KEY STATISTICAL DATA

Unit: Mtoe

SUPPLY		1973	1990	2001	2002	2010	2020	2030
TOTAL PRODUCTION		**1.40**	**3.39**	**4.10**	**3.64**	**5.64**	..	..
Coal		0.13	0.12	-	-	-	..	..
Oil		-	-	-	-	-	..	..
Gas		-	-	-	-	-	..	..
Comb. Renewables & Wastes[1]		0.64	2.48	2.76	2.84	3.79	..	..
Nuclear		-	-	-	-	-	..	..
Hydro		0.63	0.79	1.21	0.67	1.11	..	..
Geothermal		-	0.00	0.09	0.08	0.07	..	..
Solar/Wind/Other		-	0.01	0.04	0.05	0.67	..	..
TOTAL NET IMPORTS[2]		**5.69**	**14.82**	**21.56**	**22.17**	**24.34**	..	..
Coal	Exports	0.01	0.01	-	-	-	..	..
	Imports	0.28	3.00	2.97	3.47	3.07	..	..
	Net Imports	0.27	2.99	2.97	3.47	3.07	..	..
Oil	Exports	0.23	2.50	1.40	1.40	..	..	..
	Imports	6.44	14.93	18.19	17.69	17.51	..	..
	Bunkers	0.80	0.61	0.48	0.48	1.36	..	..
	Net Imports	5.42	11.82	16.32	15.81	16.15	..	..
Gas	Exports	-	-	-	-	-	..	..
	Imports	-	-	2.25	2.73	5.12	..	..
	Net Imports	-	-	2.25	2.73	5.12	..	..
Electricity	Exports	0.01	0.15	0.30	0.30	-	..	..
	Imports	0.01	0.15	0.32	0.46	-	..	..
	Net Imports	-0.00	0.00	0.02	0.16	-	..	..
TOTAL STOCK CHANGES		**0.14**	**-0.47**	**-0.22**	**0.58**	-	..	..
TOTAL SUPPLY (TPES)		**7.23**	**17.75**	**25.43**	**26.39**	**29.98**	..	..
Coal		0.51	2.76	3.19	3.48	3.07	..	..
Oil		5.45	11.71	15.87	16.38	16.15	..	..
Gas		-	-	2.25	2.73	5.12	..	..
Comb. Renewables & Wastes[1]		0.64	2.48	2.76	2.84	3.79	..	..
Nuclear		-	-	-	-	-	..	..
Hydro		0.63	0.79	1.21	0.67	1.11	..	..
Geothermal		-	0.00	0.09	0.08	0.07	..	..
Solar/Wind/Other		-	0.01	0.04	0.05	0.67	..	..
Electricity Trade[3]		-0.00	0.00	0.02	0.16	-	..	..
Shares (%)								
Coal		*7.0*	*15.5*	*12.6*	*13.2*	*10.2*	..	..
Oil		*75.4*	*66.0*	*62.4*	*62.1*	*53.9*	..	..
Gas		-	-	*8.9*	*10.3*	*17.1*	..	..
Comb. Renewables & Wastes		*8.8*	*14.0*	*10.8*	*10.7*	*12.6*	..	..
Nuclear		-	-	-	-	-	..	..
Hydro		*8.7*	*4.4*	*4.7*	*2.5*	*3.7*	..	..
Geothermal		-	-	*0.4*	*0.3*	*0.2*	..	..
Solar/Wind/Other		-	*0.1*	*0.2*	*0.2*	*2.2*	..	..
Electricity Trade		-	-	*0.1*	*0.6*	-	..	..

0 is negligible, - is nil, .. is not available

Unit: Mtoe

DEMAND

FINAL CONSUMPTION BY SECTOR

	1973	1990	2001	2002	2010	2020	2030
TFC	**6.11**	**14.00**	**20.34**	**20.78**	**23.81**	..	..
Coal	0.19	0.59	0.19	0.18	0.17	..	..
Oil	4.59	8.97	13.08	13.17	14.33	..	..
Gas	0.05	0.05	1.04	1.18	1.72	..	..
Comb. Renewables & Wastes[1]	0.58	2.33	2.41	2.46	2.55	..	..
Geothermal	-	-	0.00	0.00	-	..	..
Solar/Wind/Other	-	0.01	0.02	0.02	0.06	..	..
Electricity	0.70	2.03	3.44	3.57	4.54	..	..
Heat	-	0.03	0.16	0.20	0.44	..	..
Shares (%)							
Coal	*3.1*	*4.2*	*0.9*	*0.9*	*0.7*	..	..
Oil	*75.1*	*64.0*	*64.3*	*63.4*	*60.2*	..	..
Gas	*0.8*	*0.4*	*5.1*	*5.7*	*7.2*	..	..
Comb. Renewables & Wastes	*9.5*	*16.6*	*11.8*	*11.9*	*10.7*	..	..
Geothermal	-	-	-	-	-	..	..
Solar/Wind/Other	-	*0.1*	*0.1*	*0.1*	*0.3*	..	..
Electricity	*11.5*	*14.5*	*16.9*	*17.2*	*19.1*	..	..
Heat	-	*0.2*	*0.8*	*1.0*	*1.9*	..	..
TOTAL INDUSTRY[4]	**2.71**	**6.81**	**8.37**	**8.46**	**9.39**	..	..
Coal	0.14	0.59	0.19	0.18	0.17	..	..
Oil	1.81	3.96	4.56	4.45	4.26	..	..
Gas	0.00	-	0.83	0.92	1.20	..	..
Comb. Renewables & Wastes[1]	0.32	1.18	1.26	1.31	1.40	..	..
Geothermal	-	-	-	-	-	..	..
Solar/Wind/Other	-	-	-	-	-	..	..
Electricity	0.44	1.05	1.39	1.42	1.93	..	..
Heat	-	0.03	0.15	0.19	0.42	..	..
Shares (%)							
Coal	*5.1*	*8.7*	*2.3*	*2.1*	*1.8*	..	..
Oil	*66.9*	*58.2*	*54.4*	*52.6*	*45.4*	..	..
Gas	*0.1*	-	*9.9*	*10.8*	*12.8*	..	..
Comb. Renewables & Wastes	*11.8*	*17.3*	*15.0*	*15.5*	*14.9*	..	..
Geothermal	-	-	-	-	-	..	..
Solar/Wind/Other	-	-	-	-	-	..	..
Electricity	*16.2*	*15.4*	*16.6*	*16.7*	*20.6*	..	..
Heat	-	*0.4*	*1.8*	*2.2*	*4.5*	..	..
TRANSPORT[5]	**1.95**	**3.82**	**6.70**	**6.90**	**8.27**	..	..
TOTAL OTHER SECTORS[6]	**1.46**	**3.37**	**5.26**	**5.42**	**6.16**	..	..
Coal	0.04	0.00	-	-	-	..	..
Oil	0.87	1.21	1.86	1.86	1.85	..	..
Gas	0.05	0.05	0.21	0.26	0.52	..	..
Comb. Renewables & Wastes[1]	0.26	1.15	1.15	1.15	1.15	..	..
Geothermal	-	-	0.00	0.00	-	..	..
Solar/Wind/Other	-	0.01	0.02	0.02	0.06	..	..
Electricity	0.25	0.95	2.01	2.12	2.56	..	..
Heat	-	-	0.01	0.01	0.02	..	..
Shares (%)							
Coal	*2.4*	-	-	-	-	..	..
Oil	*59.7*	*35.9*	*35.3*	*34.4*	*30.1*	..	..
Gas	*3.2*	*1.5*	*4.0*	*4.8*	*8.4*	..	..
Comb. Renewables & Wastes	*17.9*	*34.1*	*21.9*	*21.2*	*18.7*	..	..
Geothermal	-	-	-	-	-	..	..
Solar/Wind/Other	-	*0.3*	*0.4*	*0.4*	*1.0*	..	..
Electricity	*16.8*	*28.1*	*38.3*	*39.0*	*41.6*	..	..
Heat	-	-	*0.2*	*0.2*	*0.3*	..	..

Unit: Mtoe

DEMAND

ENERGY TRANSFORMATION AND LOSSES

	1973	1990	2001	2002	2010	2020	2030
ELECTRICITY GENERATION[7]							
INPUT (Mtoe)	**1.33**	**5.10**	**7.87**	**8.42**	**10.49**	..	..
OUTPUT (Mtoe)	**0.84**	**2.44**	**3.97**	**3.93**	**5.18**	..	..
(TWh gross)	9.79	28.36	46.17	45.65	60.20	..	..
Output Shares (%)							
Coal	*3.9*	*32.1*	*29.5*	*33.3*	*21.8*	..	..
Oil	*19.2*	*33.1*	*20.2*	*25.0*	*7.9*	..	..
Gas	–	–	*15.6*	*19.8*	*33.8*	..	..
Comb. Renewables & Wastes	*2.0*	*2.4*	*3.5*	*3.8*	*3.0*	..	..
Nuclear	–	–	–	–	–	..	..
Hydro	*74.8*	*32.3*	*30.4*	*17.1*	*21.5*	..	..
Geothermal	–	*0.0*	*0.2*	*0.2*	*0.1*	..	..
Solar/Wind/Other	–	*0.0*	*0.6*	*0.8*	*11.8*	..	..
TOTAL LOSSES	**1.23**	**3.21**	**5.05**	**5.48**	**6.17**	..	..
of which:							
Electricity and Heat Generation[8]	0.49	2.63	3.74	4.29	4.69	..	..
Other Transformation	0.23	-0.38	-0.02	-0.08	–	..	..
Own Use and Losses[9]	0.51	0.96	1.34	1.26	1.48	..	..
Statistical Differences	**-0.11**	**0.53**	**0.04**	**0.14**	–	..	...

INDICATORS

	1973	1990	2001	2002	2010	2020	2030
GDP (billion 1995 US$)	57.68	98.55	132.07	132.64	169.34	..	..
Population (millions)	8.64	9.90	10.30	10.37	10.50	..	..
TPES/GDP[10]	0.13	0.18	0.19	0.20	0.18	..	..
Energy Production/TPES	0.19	0.19	0.16	0.14	0.19	..	..
Per Capita TPES[11]	0.84	1.79	2.47	2.54	2.85	..	..
Oil Supply/GDP[10]	0.09	0.12	0.12	0.12	0.10	..	..
TFC/GDP[10]	0.11	0.14	0.15	0.16	0.14	..	..
Per Capita TFC[11]	0.71	1.41	1.97	2.00	2.27	..	..
Energy-related CO_2 Emissions (Mt CO_2)[12]	16.4	39.6	59.1	63.0	66.5	..	..
CO_2 Emissions from Bunkers (Mt CO_2)	3.5	3.5	3.3	3.4	6.1	..	..

GROWTH RATES (% per year)

	73-79	79-90	90-01	01-02	02-10	10-20	20-30
TPES	5.5	5.4	3.3	3.8	1.6	..	..
Coal	-2.4	18.2	1.3	8.9	-1.5	..	..
Oil	6.1	3.8	2.8	3.2	-0.2	..	..
Gas	–	–	–	21.1	8.2	..	..
Comb. Renewables & Wastes	3.2	11.2	1.0	2.9	3.7	..	..
Nuclear	–	–	–	–	–	..	..
Hydro	7.3	-1.8	4.0	-44.4	6.5	..	..
Geothermal	–	–	36.4	-7.7	-2.4	..	..
Solar/Wind/Other	–	–	12.7	24.4	38.0	..	..
TFC	4.7	5.2	3.4	2.2	1.7	..	..
Electricity Consumption	8.5	5.3	4.9	3.8	3.1	..	..
Energy Production	4.4	5.9	1.7	-11.1	5.6	..	..
Net Oil Imports	8.1	2.9	3.0	-3.1	0.3	..	..
GDP	2.9	3.4	2.7	0.4	3.1	..	..
Growth in the TPES/GDP Ratio	2.5	2.0	0.6	3.3	-1.4	..	..
Growth in the TFC/GDP Ratio	1.8	1.7	0.7	1.7	-1.3	..	..

Please note: Rounding may cause totals to differ from the sum of the elements.

FOOTNOTES TO ENERGY BALANCES AND KEY STATISTICAL DATA

1. Comprises solid biomass, biogas and municipal waste. Data are often based on partial surveys and may not be comparable between countries.
2. Total net imports include combustible renewables and waste.
3. Total supply of electricity represents net trade. A negative number indicates that exports are greater than imports.
4. Includes non-energy use.
5. Includes less than 1% non-oil fuels.
6. Includes residential, commercial, public service and agricultural sectors.
7. Inputs to electricity generation include inputs to electricity and CHP plants. Output refers only to electricity generation.
8. Losses arising in the production of electricity and heat at public utilities and autoproducers. For non-fossil-fuel electricity generation, theoretical losses are shown based on plant efficiencies of 10% for geothermal and 100% for hydro.
9. Data on "losses" for forecast years often include large statistical differences covering differences between expected supply and demand and mostly do not reflect real expectations on transformation gains and losses.
10. Toe per thousand US dollars at 1995 prices and exchange rates.
11. Toe per person.
12. "Energy-related CO_2 emissions" specifically means CO_2 from the combustion of the fossil fuel components of TPES (*i.e.* coal and coal products, peat, crude oil and derived products and natural gas), while CO_2 emissions from the remaining components of TPES (*i.e.* electricity from hydro, other renewables and nuclear) are zero. Emissions from the combustion of biomass-derived fuels are not included, in accordance with the IPCC greenhouse gas inventory methodology. Also in accordance with the IPCC methodology, emissions from international marine and aviation bunkers are not included in national totals. Projected emissions for oil and gas are derived by calculating the ratio of emissions to energy use for 2002 and applying this factor to forecast energy supply. Future coal emissions are based on product-specific supply projections and are calculated using the IPCC/OECD emission factors and methodology.

INTERNATIONAL ENERGY AGENCY "SHARED GOALS"

The member countries* of the International Energy Agency (IEA) seek to create the conditions in which the energy sectors of their economies can make the fullest possible contribution to sustainable economic development and the well-being of their people and of the environment. In formulating energy policies, the establishment of free and open markets is a fundamental point of departure, though energy security and environmental protection need to be given particular emphasis by governments. IEA countries recognise the significance of increasing global interdependence in energy. They therefore seek to promote the effective operation of international energy markets and encourage dialogue with all participants.

In order to secure their objectives they therefore aim to create a policy framework consistent with the following goals:

1. **Diversity, efficiency and flexibility within the energy sector** are basic conditions for longer-term energy security: the fuels used within and across sectors and the sources of those fuels should be as diverse as practicable. Non-fossil fuels, particularly nuclear and hydro power, make a substantial contribution to the energy supply diversity of IEA countries as a group.

2. Energy systems should have **the ability to respond promptly and flexibly to energy emergencies**. In some cases this requires collective mechanisms and action: IEA countries co-operate through the Agency in responding jointly to oil supply emergencies.

3. **The environmentally sustainable provision and use of energy** is central to the achievement of these shared goals. Decision-makers should seek to minimise the adverse environmental impacts of energy activities, just as environmental decisions should take account of the energy consequences. Government interventions should where practicable have regard to the Polluter Pays Principle.

4. **More environmentally acceptable energy sources** need to be encouraged and developed. Clean and efficient use of fossil fuels is essential. The development of economic non-fossil sources is also a priority. A number of IEA members wish to retain and improve the nuclear option for the

* Australia, Austria, Belgium, Canada, the Czech Republic, Denmark, Finland, France, Germany, Greece, Hungary, Ireland, Italy, Japan, the Republic of Korea, Luxembourg, the Netherlands, New Zealand, Norway, Portugal, Spain, Sweden, Switzerland, Turkey, the United Kingdom, the United States.

future, at the highest available safety standards, because nuclear energy does not emit carbon dioxide. Renewable sources will also have an increasingly important contribution to make.

5. **Improved energy efficiency** can promote both environmental protection and energy security in a cost-effective manner. There are significant opportunities for greater energy efficiency at all stages of the energy cycle from production to consumption. Strong efforts by governments and all energy users are needed to realise these opportunities.

6. Continued **research, development and market deployment of new and improved energy technologies** make a critical contribution to achieving the objectives outlined above. Energy technology policies should complement broader energy policies. International co-operation in the development and dissemination of energy technologies, including industry participation and co-operation with non-member countries, should be encouraged.

7. **Undistorted energy prices** enable markets to work efficiently. Energy prices should not be held artificially below the costs of supply to promote social or industrial goals. To the extent necessary and practicable, the environmental costs of energy production and use should be reflected in prices.

8. **Free and open trade** and a secure framework for investment contribute to efficient energy markets and energy security. Distortions to energy trade and investment should be avoided.

9. **Co-operation among all energy market participants** helps to improve information and understanding, and encourage the development of efficient, environmentally acceptable and flexible energy systems and markets worldwide. These are needed to help promote the investment, trade and confidence necessary to achieve global energy security and environmental objectives.

(The *Shared Goals* were adopted by IEA Ministers at their 4 June 1993 meeting in Paris.)

GLOSSARY AND LIST OF ABBREVIATIONS

In this report, abbreviations are substituted for a number of terms used within the International Energy Agency. While these terms generally have been written out on first mention and subsequently abbreviated, this glossary provides a quick and central reference for many of the abbreviations used.

bcm	billion cubic metres.
cm	cubic metre.
CAE	power purchase agreement.
CCGT	combined-cycle gas turbine.
CDM	clean development mechanism.
CHP	combined production of heat and power; sometimes, when referring to industrial CHP, the term "co-generation" is used.
DGGE	General Directorate for Geology and Energy.
E4	Energy Efficiency and Endogenous Energy programme.
EC	European Commission.
EGREP	Entidade Gestora de Reservas Estratégicas de Produtos Petrolíferos.
ERSE	Energy Services Regulatory Authority.
EU	European Union.
EU-ETS	EU Emissions Trading Scheme.
GDP	gross domestic product.
GHG	greenhouse gases (see footnote 2).
GW	gigawatt, or 1 watt $\times 10^9$.
GWh	gigawatt-hour = 1 gigawatt $\times$ 1 hour.
HFC	hydrofluorocarbons.
HVAC	heating, ventilating and air-conditioning.

IEA	International Energy Agency.
IEP	International Energy Program.
IES	Independent Electricity System.
INETI	National Institute for Engineering and Industrial Technology.
IPP	independent power producer.
JI	joint implementation.
km^2	square kilometre.
ktoe	thousand tonnes of oil equivalent; see toe.
kW	kilowatt, or 1 watt $\times 10^3$.
kWh	kilowatt-hour = 1 kilowatt $\times$ one hour.
kV	kilovolt, or 1 volt $\times 10^3$.
LNG	liquefied natural gas.
LPG	liquefied petroleum gas.
m	metre.
m^2	square metre.
m^3	cubic metre.
MAPE	Measure to Support the Harnessing of the Energy Potential and Rationalisation of Consumption.
mcm	million cubic metres.
MIBEL	Iberian Electricity Market.
Mt	million tonnes.
Mtoe	million tonnes of oil equivalent; see toe.
MW	megawatt, or 1 watt $\times 10^6$.
MWh	megawatt-hour = 1 megawatt $\times$ one hour.
NAP	National Allocation Plan.
NBES	non-binding system.
NGO	non-governmental organisation.
NMVOCs	non-methane volatile organic compounds.
NO_x	nitrogen oxide.

OECD	Organisation for Economic Co-operation and Development.
OPE	Economic Operational Plan.
PES	public electricity system.
PNAC	Climate Change National Programme.
PRIME	Programme of Incentives to the Modernisation of the Economy.
RCCTE	Regulations on the Characteristic of the Thermal Behaviour of Buildings.
R&D	research and development, especially in energy technology; may include the demonstration and dissemination phases as well.
REN	Rede Electrica Nacional (electricity transmission system operator).
RGCE	Management Regulations for Energy Consumption.
RSECE	Regulations on Energy Systems and HVAC in Buildings.
SO_2	sulphur dioxide.
TFC	total final consumption of energy.
Toe	tonnes of oil equivalent, defined as 10^7 kcal.
TPA	third-party access.
TPES	total primary energy supply.
TSO	transmission system operator.
TW	terawatt, or 1 watt $\times$ 10^{12}.
TWh	terawatt-hour =1 terawatt $\times$ one hour.
VAT	value added tax.
VOCs	volatile organic compounds.

IEA PUBLICATIONS, 9, rue de la Fédération, 75739 PARIS CEDEX 15
PRINTED IN FRANCE BY STEDI
(61 04 21 1P1) ISBN 92-64-1079-91 - 2004